Heavy Ion Accelerator Symposium 2015

International Nuclear Structure Conference in Remembrance of George Dracoulis

Canberra, Australia
14-18 September 2015

Editors:

M. W. Reed

E. C. Simpson
A. J. Mitchell

ISBN: 978-1-5108-3016-5

Printed from e-media with permission by:

Curran Associates, Inc.
57 Morehouse Lane
Red Hook, NY 12571

Some format issues inherent in the e-media version may also appear in this print version.

Printed by Curran Associates, Inc. (2016)

For additional information, please contact EDP Sciences – Web of Conferences
at the address below.

EDP Sciences – Web of Conferences
17, Avenue du Hoggar
Parc d'Activité de Courtabœuf
BP 112
F-91944 Les Ulis Cedex A
France

Phone: +33 (0) 1 69 18 75 75
Fax: +33 (0) 1 69 28 84 91

contact-edps@webofconferences.org

Additional copies of this publication are available from:

Curran Associates, Inc.
57 Morehouse Lane
Red Hook, NY 12571 USA
Phone: 845-758-0400
Fax: 845-758-2633
Email: curran@proceedings.com
Web: www.proceedings.com

TABLE OF CONTENTS

Preface

In September 2015 the annual Heavy Ion Accelerator Symposium took place at the Australian National University (ANU). On this occasion the special purpose of the Symposium was to celebrate the life and achievements of Professor George Dracoulis, who sadly passed away in June 2014. George was a meticulous researcher who strove for perfection. He had high standards, not just for himself, but for his colleagues, post-docs and students. As one of his closest colleagues and collaborators put it 'if George had measured it, then you knew it was right'. George joined the ANU Department of Nuclear Physics around the time of the commissioning of the 14UD Pelletron accelerator and played a major role in establishing the ANU laboratory as one of the most respected nuclear-physics laboratories in the world, a legacy that continues today. George provided leadership as Head of Department from 1992 until his retirement in 2009. The Symposium was designed to not only look back at George's legacy, but also to look forward and discuss current and future research growing out of those achievements. These proceedings give a representation of these goals; a number of articles are directly related to fields George was most prominent in, and others build upon these topics, looking to the future.

Matthew Reed

Editor-in-chief

Australian National University

19 August 2016

Dr Matthew Reed
Research Fellow

Department of Nuclear Physics
Research School of Physics & Engineering
College of Physical & Mathematical Sciences

T: +61 2 6125 200609
F: +61 2 6125 0748
E: matthew.reed@anu.edu.au
www.physics@anu.edu.au/nuclear

Nuclear Physics Building 57
West end of Garran Road
Canberra ACT 2601 Australia

Subject: Statement of Peer-Review for the publication of HIAS15 proceedings

Dear publishers,

All contributions to the Heavy-Ion Accelerator Symposium (HIAS) 2015 proceedings have been peer-reviewed by anonymous international experts (one referee per contribution) according to international standards and practices. The referees were chosen by the editorial board (M. W. Reed, E. C. Simpson and A. J. Mitchell) nominated by G. J. Lane, Chair of HIAS15. The authors were given a chance to reply to the referee.

Sincerely yours,

Dr. Matthew Reed

Dr Edward Simpson

Dr Alan John Mitchell

EPJ Web of Conferences **123**, 01001 (2016)
Heavy Ion Accelerator Symposium 2015

DOI: 10.1051/epjconf/201612301001

High-K isomers: some of the questions

P. M. Walker[1,a]

[1] *Department of Physics, University of Surrey, Guildford GU2 7XH, United Kingdom*

Abstract. High-K isomers exemplify the coexistence of individual-particle and collective motion in atomic nuclei. Here, the topic is briefly outlined, and some open questions are discussed. These include violations of the K quantum number; the high-spin limit to K isomerism; the fission stability of K isomers; possibilities for manipulation and control of K-isomer decay rates; and access to K isomers in neutron-rich nuclei.

1 Introduction

This conference celebrates the scientific life and work of George Dracoulis, who sadly passed away in 2014. I was privileged to be his first PhD student. During my PhD, we made an experimental study of high-spin states in ^{172}Hf [1, 2], and we thus embarked on a life-long journey exploring long-lived, excited states in deformed nuclei, the so-called K isomers. Amongst our many joint publications are a 1999 Nature review of isomers [3]; a 2001 review focussed on high-K isomers [4]; and finally an in-depth isomer review, soon to be published [5], that Dracoulis worked on until shortly before he died. He also published other related reviews, notably his recent Nobel symposium paper [6], and a new and comprehensive K-isomer tabulation [7].

The present paper introduces a selection of open questions relating to K isomers, where much research remains to be done. As we develop our understanding and push towards the outer reaches of the nuclear chart, one of the experimental issues that needs to be kept in mind was stated by Dracoulis [6]: "There is always a problem to be aware of with the study of isomers ... and that is that the popular techniques for their identification can be compromised if the lifetimes become very long, as might occur, paradoxically, in the more interesting cases."

First, though, some terminology needs a brief explanation. The K quantum number represents the projection of the nuclear angular momentum onto its deformation axis (symmetry axis) with the deformed shape being prolate in the vast majority of cases. High K values can be made by broken-pair, deformation-aligned excitations, with each broken pair increasing the number of quasiparticles (unpaired nucleons) by two units. Electromagnetic transitions from (multi-)quasiparticle states are called "K forbidden" if the change in K exceeds the multipole order of the transition, i.e. if $\Delta K > \lambda$. However, due to K-mixing mechanisms (rotational, vibrational or statistical [4]) such transition are hindered, rather than strictly forbidden. The degree of forbiddenness is defined as $\nu = \Delta K - \lambda$, and the reduced hindrance is expressed as $f_\nu = (F_W)^{1/\nu}$, where F_W is the Weisskopf hindrance factor [4, 7]. In this way, f_ν represents the hindrance per degree of K forbiddenness, giving some measure of the effect of K-mixing processes, with large f_ν values corresponding to little K mixing.

[a] e-mail: p.walker@surrey.ac.uk

2 Erosion of the K quantum number

There is no strict definition of the half-life needed for a nuclear excited state to be termed an "isomer". Nevertheless, it is clear that short half-lives generally correspond to low f_ν values. With germanium γ-ray detectors, it becomes difficult to determine half-lives of less than about 10 ns. One such case from the early work of Dracoulis and Walker [1, 2], alluded to above, is the half-life of the two-quasiproton, $K^\pi = 6^+$ isomer in ^{172}Hf. The curve fitting necessary to obtain the 5 ns value is illustrated in Fig. 1.

Perhaps surprisingly, after nearly four decades, this ^{172}Hf isomer is still at the edge of accessibility, with regard to having a measureable half-life: the corresponding state in the lighter adjacent even-even isotope, ^{170}Hf, only has a half-life limit, measured to be <5 ns [9]. The advent of fast-timing $LaBr_3$, γ-ray detectors (see, for example, Ref. [10]) may well remove this impasse, and open the door to a range of shorter-lived K isomers.

Nevertheless, as of now, there are five even-even hafnium isotopes with $K^\pi = 6^+$ isomers that have known half-lives, as given in Table 1. The most recently determined value is for ^{180}Hf, now listed in the 2015 review of Kondev *et al.* [7]. As seen in Table 1, the half-lives range over almost four orders of magnitude, and the f_ν values for the $E2$ decay branches vary by a factor of six. Clues to the observed behaviour come from inspection of the dependence of f_ν on the product of the valence nucleon numbers, N_pN_n, revealing a strong correlation [11, 12], as illustrated in Fig. 2. (Note that the new f_ν value for ^{180}Hf is in good

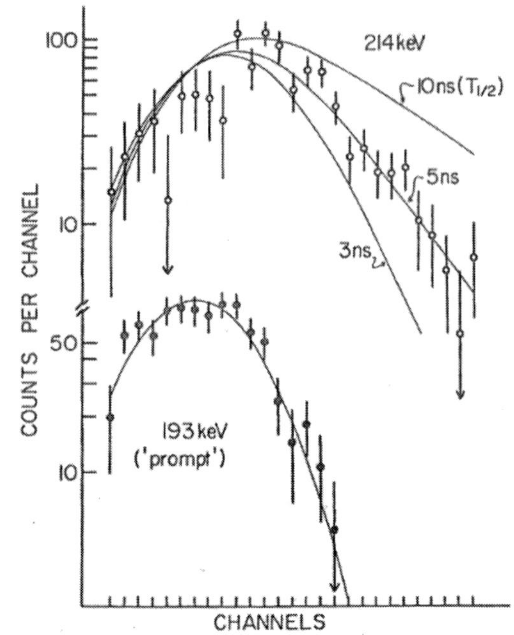

Figure 1. Illustration of the Gaussian-plus-exponential curve fitting needed to obtain the 5 ns half-life of the $K^\pi = 6^+$ isomer in ^{172}Hf, for 214 keV events (open circles) shown as a function of time, from Ref. [2]. The Gaussian curve for the 193 keV transition (filled circles) represents the "prompt" lineshape. Later measurements [8] obtained 4.8±0.4 ns for the half-life of the 6^+ state.

accord.) Qualitatively, the behaviour seems to be reasonably simple: small values of N_pN_n correspond to weak collectivity, which can reasonably be associated with poor K conservation, hence small f_ν values. However, a quantitative understanding of this relationship remains elusive. In contrast, a recent analysis by Chen *et al.* [13] of $K^\pi = 6^+$ isomers in the $N = 104$ isotones identifies mixing with the $I^\pi = 6^+$ member of the $K = 2$, γ-vibrational band as being a key determinant of the f_ν value. Whether this kind of analysis is also able to explain the hafnium f_ν values in Table 1 and, indeed, to predict as-yet unmeasured half-lives, remains to be evaluated.

Extension of the f_ν systematics from two-quasiparticle to multi-quasiparticle isomers is discussed elsewhere [4, 5, 7, 14]. For multi-quasiparticle states, it is evident that the level density is an important variable. Multi-quasiparticle states that are further above yrast (i.e. relative to the energy of the lowest-lying rotational state of the same spin) tend to have less-hindered decays and thus have shorter half-lives.

3 High-spin limit to K

High-K isomers are known up to spin $57/2\hbar$, which is found for a nine-quasiparticle, $T_{1/2}$ = 22 ns state in ^{175}Hf [7, 15]. It might be thought that high spin would destroy

Table 1. Properties of $K^\pi = 6^+$ isomers in hafnium ($Z = 72$) isotopes, including reduced-hindrance (f_ν) values for $E2$, $\Delta K = 6$ decay branches. Data are from Ref. [7].

nuclide	E (keV)	$T_{1/2}$	f_ν $(E2)$	N_pN_n
^{170}Hf	1773	<5 ns	<4	160
^{172}Hf	1685	5 ns	10	180
^{174}Hf	1549	140 ns	17	200
^{176}Hf	1333	10 μs	42	220
^{178}Hf	1554	80 ns	16	200
^{180}Hf	1703	3 ns	7	180

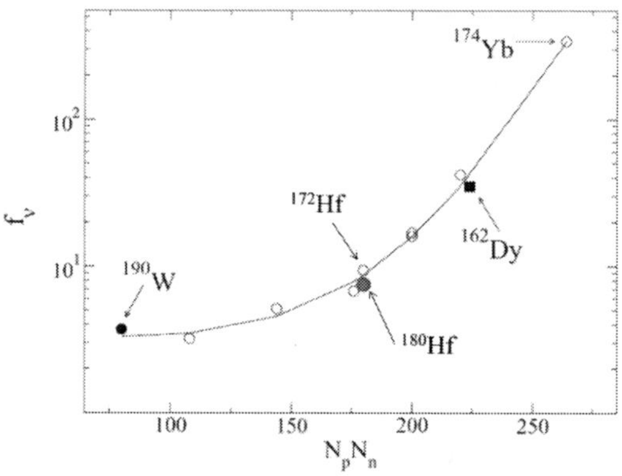

Figure 2. Reduced hindrance, f_ν, as a function of the product of the valence nucleon numbers, N_pN_n, for $\Delta K \geq 6$, $E2$ decays from two-quasiparticle isomers in even-even nuclei in the $A \approx 160-190$ deformed region. The figure is adapted from Ref. [12], with the inclusion of the ^{180}Hf data point [7]. The line through the data points is to guide the eye.

the K quantum number, on account of Coriolis K mixing. However, the K isomers themselves derive their angular momentum from individual nucleon orbits, rather than collective rotation, and it seems that the angular-momentum limit has not been reached. Indeed, the limit is most likely experimental, in the sense that the theoretically favoured nuclei (in the neutron-rich hafnium region [16]) cannot be synthesised at high spin, at least not yet in a way that enables their experimental study (though see also Section 6). According to calculations [16], $K > 40$ isomers will be yrast in neutron-rich hafnium isotopes, and the eventual limit would come from competition with coexisting rotation-aligned, oblate-deformed states.

4 Fissioning K isomers

One of the limits to nuclear binding is that of spontaneous fission. It can be argued that there is, as yet, no case where a K isomer has been conclusively established to fission [17, 18], and that, more generally, K isomers can provide extra stability for the heaviest elements [19]. The most recent example of resistance to fission is in ^{254}Rf [20],

 DOI: 10.1051/epjconf/201612301001

where a proposed four-quasiparticle isomer has a half-life of 250 μs, ten times longer than its spontaneously fissioning ground state, and no isomeric fission branch could be detected. The outstanding experimental challenge is therefore to measure the degree to which fission is inhibited from K isomers, i.e. to determine unambiguously a K-isomer fission probability. The corresponding theoretical description is also challenging.

5 Isomer manipulation

The high-energy-density storage capability of isomers is of long-standing interest [3, 21], and ways to release that energy have been sought. The photo-activation of ^{180m}Ta, the sole naturally occurring isomer, has been relatively well studied [22, 23], but the need for $\approx$1 MeV photons to release 75 keV of isomer energy mitigates against practical applications.

Initial studies of ^{178m2}Hf, the 31 year, $K^\pi = 16^+$ isomer of ^{178}Hf, at 2.4 MeV excitation energy, showed great promise for its energy release using <100 keV photons incident on radioactive targets of ^{178m2}Hf [24]. However, much lower cross-section limits were later set [25, 26]. Now there are again positive results reported [27], but this time at the lower cross-section values. This remains a controversial area of work.

Other avenues to induce isomer decay electromagnetically include Coulomb excitation with radioactive beams [28], nuclear superradiance [29], and nuclear excitation by electron capture [30, 31], though the latter remains unobserved. The field remains open for innovation and exploitation.

6 K isomers in neutron-rich nuclei

It is well known that a large number of nuclides remain to be discovered on the neutron-rich side of β stability, and an even larger number await having their structural properties measured. One of the problems is that they are typically produced in non-selective reactions and with tiny cross sections, in relativistic projectile fragmentation and fission reactions, so that the γ rays emitted at the target position cannot in general be related to specific nuclides. Microsecond isomers offer a selective route to structure information, since they can be transported through mass/charge separators, identified on an ion-by-ion basis, and have their γ-ray emissions time-correlated with the identified ions. Such isomers are now giving access to excited states in, for example, the doubly-mid-shell rare-earth deformed region [32], which is a region of both astrophysics and nuclear structure interest, relating to r-process nucleosynthesis, possible subshell gaps, and high-order (β_6) deformation. The growth in this knowledge is strongly related to the new generation of radioactive-beam facilities.

For long half-lives, the time correlation between identified ions and emitted γ rays is lost. An alternative approach is through the exploitation of heavy-ion storage rings, where single-ion sensitivity can be achieved. For example, a long-lived ($T_{1/2} \approx 12$ minutes) isomer has been discovered in ^{184}Hf, at an excitation energy of 2.5 MeV [33]. The challenge is now to measure its decay radiations, and establish its spectroscopic properties.

7 Outlook

Selected aspects of K isomerism have been discussed, with a focus on some of the open questions. These are interrelated. For example, an understanding of the breakdown of K isomerism through K mixing might lead to a breakthrough in isomer manipulation; and studying K isomers at very high spin is related to the problem of limited access to $A \approx 190$ neutron-rich nuclides. Perhaps one of the key future developments will be the production of high-quality isomeric beams [34], which could be used to initiate novel nuclear reactions. Also, the production of intense neutron-rich beams will open up that elusive side of the nuclear landscape at high spin. At the same time, there is the prospect that isomers will become increasingly important, through their extended half-lives, when approaching the limits of nuclear stability in heavy and superheavy nuclei.

Acknowledgements

This article is dedicated to the memory of George Dracoulis. The research is supported by the United Kingdom Science and Technology Facilities Council under grant No. ST/L005743/1.

References

[1] P. M. Walker, G. D. Dracoulis, A. Johnston and J. R. Leigh, Nucl. Phys. A **293**, 481 (1977).

[2] P. M. Walker, PhD Thesis *Australian National University* unpublished (1978).

[3] P. M. Walker and G. D. Dracoulis, Nature **399**, 35 (1999).

[4] P. M. Walker and G. D. Dracoulis, Hyperfine Interact. **135**, 83 (2001).

[5] G. D. Dracoulis, P. M. Walker and F. G. Kondev, Rep. Prog. Phys. **79**, 076301 (2016).

[6] G. D. Dracoulis, Phys. Scr. T **152**, 014015 (2103).

[7] F. G. Kondev, G. D. Dracoulis and T. Kibédi, At. Data Nucl. Data Tables **103-104**, 50 (2015).

[8] P. M. Walker, D. Ward, O. Häusser, H. R. Andrews, and T. Faestermann, Nucl. Phys. A **349**, 1 (1980).

[9] D. M. Cullen *et al.*, Phys. Rev. C **60**, 057303 (1999).

[10] F. Browne *et al.*, Phys. Lett. B **750**, 448 (2015).

[11] P. M. Walker, J. Phys. G **16**, L233 (1990).

[12] T. P. D. Swan *et al.*, Phys. Rev. C **83**, 034322 (2011).

[13] F. Q. Chen, Y. Sun, P. M. Walker, G. D. Dracoulis, Y. R. Shimizu, and J. A. Sheikh, J. Phys. G **40**, 015101 (2013).

[14] P. M. Walker and F. R. Xu, Phys. Scr. **91**, 013010 (2016).

[15] N. L. Gjørup, M. A. Bentley, B. Fabricius, A. Holm, J. F. Sharpey-Schafer, G. Sletten, and P. M. Walker, Z. Phys. A **337**, 353 (1990).

DOI: 10.1051/epjconf/201612301001

[16] F. R. Xu, P. M. Walker, and R. Wyss, Phys. Rev. C **62**, 014301 (2000).

[17] H. L. Liu, F. R. Xu, Y. Sun, P. M. Walker, and R. Wyss, Eur. Phys. J. A **47**, 135 (2011).

[18] P. M. Walker, F. R. Xu, H. L. Liu, and Y. Sun, J. Phys. G **39**, 105106 (2012).

[19] F. R. Xu, E. G. Zhao, R. Wyss, and P. M. Walker, Phys. Rev. Lett. **92**, 252501 (2004).

[20] H. M. David *et al.*, Phys. Rev. Lett. **115**, 132502 (2015).

[21] J. J. Carroll, Nucl. Inst. Meth. Phys. Res. B **261**, 960 (2007).

[22] D. Belic, *et al.*, Phys. Rev. Lett. **83**, 5242 (1999); Phys. Rev. C **65**, 035801 (2002).

[23] P. M. Walker, G. D. Dracoulis, and J. J. Carroll, Phys. Rev. C **64**, 061302(R) (2001).

[24] C. B. Collins *et al.*, Phys. Rev. Lett. **82**, 695 (1999).

[25] I. Ahmad *et al.*, Phys. Rev. C **71**, 024311 (2005).

[26] J. J. Carroll *et al.*, Phys. Lett. B **679**, 203 (2009).

[27] V. I. Kirischuk, V. A. Ageev, A. M. Dovbnya, S. S. Kandybei, and Yu. M. Ranyuk, Phys. Lett. B **750**, 89 (2015).

[28] I. Stefanescu *et al.*, Phys. Rev. Lett. **98**, 122701 (2007).

[29] R. Röhlsberger, K. Schlage, B. Sahoo, S. Couet and R. Rüffer, Science **328**, 1248 (2010).

[30] A. Pálffy, J. Evers and C. H. Keitel, Phys. Rev. Lett. **99**, 172502 (2007).

[31] G. Gosselin, V. Méot and P. Morel, Phys. Rev. C **76**, 044611 (2007).

[32] Z. Patel *et al.*, Phys. Rev. Lett. **113**, 262502 (2014).

[33] M. W. Reed *et al.*, Phys. Rev. Lett. **105**, 172501 (2010).

[34] P. M. Walker, Yu. A. Litvinov and H. Geissel, Int. J. Mass Spect. **349-350**, 247 (2013).

 DOI: 10.1051/epjconf/201612301002

Multi-quasiparticle high-K isomeric states in deformed nuclei

F. R. Xu[1,2], Y. Shi[1], H. L. Liu[1], W. Y. Liang[1], P. M. Walker[3] and G. D. Dracoulis[4]

[1] *State Key Laboratory of Nuclear Physics and Technology, School of Physics, Peking University, Beijing 100871, China*
[2] *State Key Laboratory of Theoretical Physics, Institute of Theoretical Physics, Chinese Academy of Sciences, Beijing 100190, China*
[3] *Department of Physics, University of Surrey, Guildford GU2 7XH, United Kingdom*
[4] *Department of Nuclear Physics, Research School of Physics and Engineering, The Australian National University, Canberra, ACT 2601, Australia*

Abstract. In the past years, we have made many theoretical investigations on multi-quasiparticle high-K isomeric states. A deformation-pairing-configuration self-consistent calculation has been developed by calculating a configuration-constrained multi-quasiparticle potential energy surface (PES). The specific single-particle orbits that define the high-K configuration are identified and tracked (adiabatically blocked) by calculating the average Nilsson numbers. The deformed Woods-Saxon potential was taken to give single-particle orbits. The configuration-constrained PES takes into account the shape polarization effect. Such calculations give good results on excitation energies, deformations and other structure information about multi-quasiparticle high-K isomeric states. Many different mass regions have been investigated.

1 Introduction

Atomic nuclei can be excited by breaking paired nucleons [1–4]. If the unpaired nucleons couple to a high angular momentum (usually resulting in a large angular momentum projection onto the symmetry axis of the deformed nucleus, named K), the excited state can be an isomer due to the forbiddenness of electromagnetic decays from high-K to low-K states [1, 2]. For example, the $K^\pi = 8^-$ isomers with the configuration $\pi^2\{7/2^+[404] \otimes 9/2^-[514]\}$ systematically occur in $^{170-184}$Hf, with half-lives that range from nanoseconds to hours. The well-known 31-yr $K^\pi = 16^+$ isomer at 2.4 MeV in ^{178}Hf is still attracting interest for possible applications. An abundance of high-K isomers have been observed across the entire chart of nuclides [1, 2] including superheavy (see Ref. [5] and references therein) and drip-line mass regions [6, 7].

Theoretically, previous calculations were usually performed assuming that isomers have the same deformation as the ground state of the nucleus (see, e.g., [8]). However, high-Ω orbits (Ω is defined as the nucleon angular momentum projection onto the symmetry axis of the deformed nucleus) usually have a strong deformation-driving force, that would make the isomer shape deviate from that of the ground state. Shape polarization is more obvious in soft nuclei [9–12]. Furthermore, the pairing is another crucial factor that drastically affects the excitation energy of multi-quasiparticle states. The reduction of pairing in broken-pair multi-quasiparticle states leads to different pairing energies from the ground state. Configuration-constrained PES calculation takes into account these aspects in a self-consistent manner with respect to deformation and pairing [3, 13, 14].

2 Model

We start from the non-axial deformed Woods-Saxon (WS) potential and use the universal potential parameters [15], with the monopole pairing strength G determined by the average gap method [16]. The quadrupole pairing can also be included [17] but its effect on the excitation energies of multi-quasiparticle states is small [18]. Due to nucleon pair(s) broken in multi-quasiparticle states, the pairing correlation is reduced remarkably in high-K states. To avoid the spurious phase transition encountered in the BCS approach, we approximate particle number projection by means of the Lipkin-Nogami (LN) pairing method. In the macroscopic-microscopic model, the total energy of a state contains a macroscopic part that is calculated by the liquid-drop model and a microscopic part that can be obtained by the Strutinsky shell correction. This leads to the energy of a configuration that is given by [3]

$$\begin{aligned} E_{LN} &= \sum_{j=1}^{S} e_{k_j} + \sum_{k\neq k_j} 2V_k^2 e_k - \frac{\Delta^2}{G} - G\sum_{k\neq k_j} V_k^4 \\ &\quad + G\frac{N-S}{2} - 4\lambda_2 \sum_{k\neq k_j} (U_k V_k)^2, \end{aligned} \tag{1}$$

where e_k is the energy of the k-th single-particle orbit, and V_k^2 (or U_k^2) is the occupation (or unoccupation) probability of the orbit in the BCS wave function. G is the pairing strength, and Δ is the pairing gap. S is the seniority

 DOI: 10.1051/epjconf/201612301002

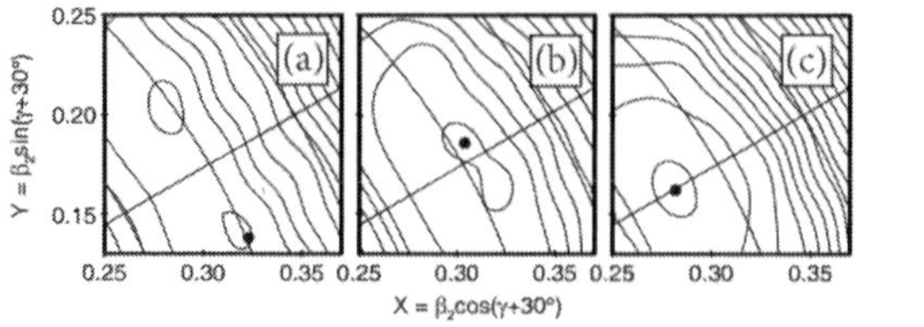

Figure 1. Calculated potential energy surfaces for the lowest-energy two-quasiparticle states in (a) ^{104}Zr ($K^\pi = 5^-, \nu\frac{5}{2}^-[532] \otimes \frac{5}{2}^+[413]$), (b) ^{108}Zr ($K^\pi = 4^-, \nu\frac{7}{2}^-[523] \otimes \frac{1}{2}^+[411]$), and (c) ^{110}Zr ($K^\pi = 6^-, \nu\frac{5}{2}^+[402] \otimes \frac{7}{2}^-[523]$), as a function of quadrupole deformation parameters β_2 and γ, with each value minimized with respect to β_4. The energy interval between contours is 200 keV.

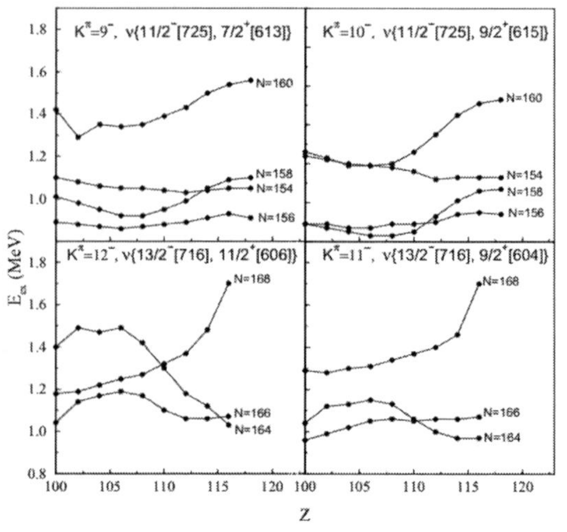

Figure 2. Calculated excitation energies for the predicted two-quasineutron high-K states of superheavy nuclei.

of the given type of nucleon, i.e., the number of unpaired protons or neutrons (indicated by k_j), and N is the number of protons or neutrons. λ_2 is a Lagrange multiplier that brings the second-order (ΔN^2) correction to the fluctuation of the particle number in the BCS wave function. Adiabatic blockings are achieved by tracking the specific single-particle orbits by calculating their average Nilsson numbers that evolve smoothly with changing deformation [3].

The potential energy surface (PES) is calculated in a lattice of quadrupole deformations (β_2, γ) with a hexadecapole β_4 variation [3]. The β_4 deformation is determined by minimizing the energy at each (β_2, γ) point. The γ deformation plays an important role in the description of nuclear collective rotation [17, 19, 20] and high-K excitations [21–25]. The γ deformation results in K-mixing (K is a conserved number for axially symmetric shape [26–28]). In heavy mass regions, higher-order deformations, e.g., β_6 and β_8, or reflection-asymmetric octupole deformation β_3 can appear [29–31]. Oblate isomers have also been predicted [32]. The pairing strength is readjusted by including mean-field and blocking effects [33]. The readjustment is crucial to reproduce experimental excitation energies.

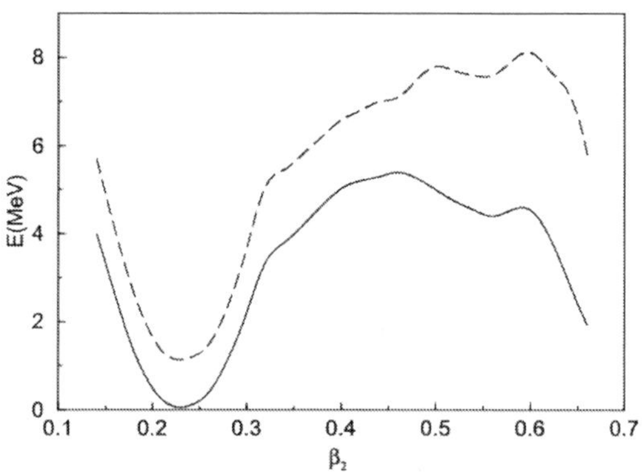

Figure 3. Calculated potential energies versus β_2 deformation in ^{256}Fm. The solid and dashed lines represent the ground state and the $\pi\frac{7}{2}^+[633] \otimes \frac{7}{2}^-[514]$ isomer, respectively. At each β_2 point, the energy has been minimized with respect to the γ and β_4 deformations.

3 Calculations

We have investigated multi-quasiparticle high-K states in different mass regions. Furthermore, PES calculations give excitation energies, deformations, g-factors, and deformation softness. The configurations can be analysed by comparing with experimental observations. Figure 1 shows configuration-constrained PES's for the lowest-energy two-quasiparticle high-K states in neutron-rich zirconium isotopes [34]. It can be seen that the calculated PES's show energy minima (giving the deformations of the states) and situations of deformation softnesses. The information is useful for analyses of the structure of a nucleus. Both triaxial deformation and softness can lead to K mixing and hence affect the decay properties of the multi-quasiparticle state.

It is interesting that high-K isomers can have longer lifetimes than corresponding ground states [35]. This is due to the K selection rule in electromagnetic transitions. We have made a systematic investigation of possible high-K isomers in superheavy nuclei [5]. Figure 2 predicts possible two-quasineutron states in superheavy nuclei. The calculated excitation energies are low, around 1.0 – 1.5 MeV. The low-lying high-K states should be isomers that can be observed experimentally. Both calculations [5, 38] and experiments [35–37] show that high-K isomerism can enhance the stability of superheavy nuclei, that is important for the experimental study of the heaviest nuclei. Figure 3 displays the calculated energy curves as a function of deformation for the ground state and a

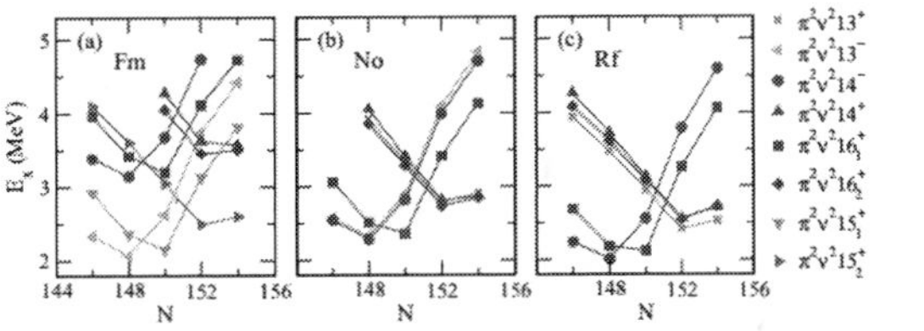

Figure 4. Predicted excitation energies for four-quasiparticle high-K states in Fm, No, and Rf nuclei. Configurations are $\pi^2\nu^2 13^+\{\pi 1/2^-[521], \pi 9/2^+[624], \nu 7/2^+[613], \nu 9/2^-[734]\}$, $\pi^2\nu^2 13^-\{\pi 7/2^+[633], \pi 7/2^-[514], \nu 5/2^+[622], \nu 7/2^+[624]\}$, $\pi^2\nu^2 14^-\{\pi 7/2^-[514], \pi 9/2^+[624], \nu 5/2^+[622], \nu 7/2^+[624]\}$, $\pi^2\nu^2 14^+\{\pi 7/2^-[514], \pi 9/2^+[624], \nu 3/2^+[622], \nu 9/2^-[734]\}$, $\pi^2\nu^2 16_1^+\{\pi 7/2^-[514], \pi 9/2^+[624], \nu 7/2^+[624], \nu 9/2^-[734]\}$, $\pi^2\nu^2 16_2^+\{\pi 7/2^-[514], \pi 9/2^+[624], \nu 7/2^+[613], \nu 9/2^-[734]\}$, $\pi^2\nu^2 15_1^+\{\pi 7/2^+[633], \pi 7/2^-[514], \nu 7/2^+[624], \nu 9/2^-[734]\}$, $\pi^2\nu^2 15_2^+\{\pi 7/2^+[633], \pi 7/2^-[514], \nu 7/2^+[613], \nu 9/2^-[734]\}$.

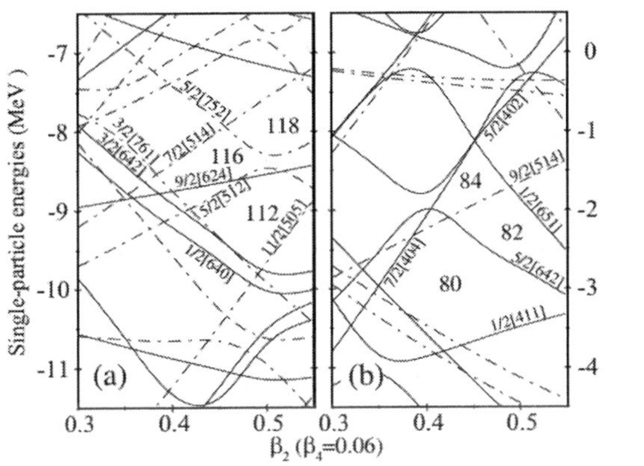

Figure 5. Neutron (a) and proton (b) single-particle levels plotted against β_2 deformation ($\beta_4 = 0.06, \gamma = 0°$).

$K^\pi = 7^-\ \pi 7/2^+[633] \otimes 7/2^-[514]$ isomer in ^{256}Fm. We see that the isomer has a higher and wider barrier than the ground state. It implies that the isomer should be more stable against fission which is the main decay mode of the heaviest nuclei. Figure 4 predicts some low-lying four-quasiparticle high-K states in Fm, No and Rf isotopes. Experiments have observed two-quasiparticle and four-quasiparticle isomers in ^{254}No [36]. Our configuration-constrained PES calculations give good descriptions of the experimental excitation energies and configuration assignments [5, 38]. It has also been found that high-spin isomerism can enhance the stability of drip-line nuclei [6], which may extend the borders of the nuclide chart.

High-K states at extremely elongated shapes present another interesting topic related to nuclear isomerism. A large axially-symmetric prolate deformation would provide a favoured condition to preserve the intrinsic angular momentum K, and hence enhance the stability of the metastable state. Experiments have observed many superdeformed (SD) rotational bands. The most well-known mass region for the SD rotation is around $A \sim 190$. It is an interesting open question whether multi-quasiparticle high-K states exist at superdeformation and can be both populated and detected in experiment. Figure 5 shows that there are many high-Ω single-particle orbits in the SD range of $\beta_2 \approx 0.4 - 0.5$. If SD high-$K$ states are located at low energies (relative to SD rotational bands), they would be populated.

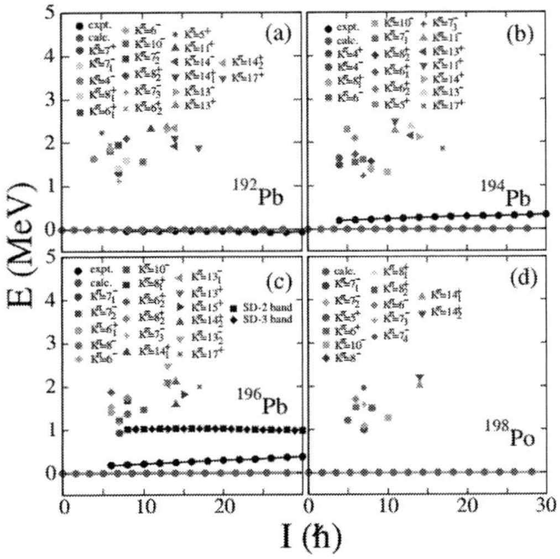

Figure 6. Excitation energies relative to the calculated yrast SD band. Experimental data are in black. A variety of multi-qp states are shown for ^{192}Pb [(a)], ^{194}Pb [(b)], ^{196}Pb [(c)], and ^{198}Po [(d)].

With the configuration-constrained PES method, we have performed systematic calculations to search for possible SD multi-quasiparticle high-K states in the mass $A \sim 190$ region [39]. Figure 6 gives predicted SD two-quasiparticle high-K states. Their excitation energies are about 1.0 – 2.0 MeV higher than the calculated yrast SD rotational band. Figure 7 shows a scheme for electromagnetic transitions from possible SD high-K bands, which may provide guidance for future experiments.

4 Summary

We have performed deformation-pairing self-consistent configuration-constrained potential-energy-surface calculations for multi-quasiparticle high-K states in different mass regions. The calculations show that high-K states exist systematically in the superheavy mass region. The calculated excitation energies of high-K states can compete with those of collective rotational states. Experimentally several high-K isomers have been identified in the superheavy region. The present calculations can well reproduce the experimental data. We have also predicted superdeformed high-K states in the $A \sim 190$ region where superdeformed rotational bands have been observed. The calculated excitation energies are about

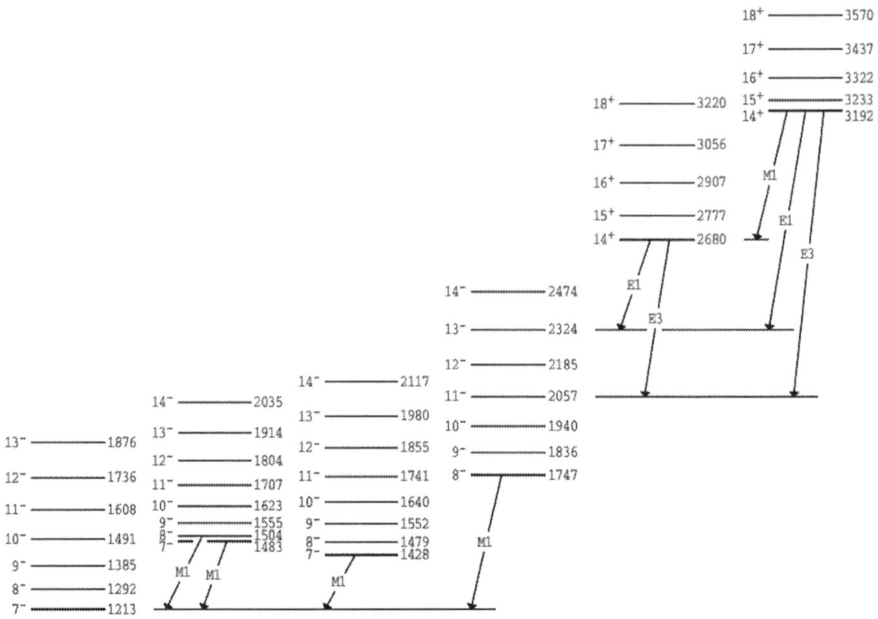

Figure 7. Notional decay scheme of calculated high-K SD bands in ^{196}Pb, showing the main decay paths expected between bands. Energies are in keV, relative to the SD minimum.

1.0 – 2.0 MeV above the observed yrast superdeformed rotational band. The predictions can provide useful guidance for experiments. The collective rotations of multi-quasiparticle high-K states have also been calculated by the configuration-constrained total Routhian surface method [40, 41].

Acknowledgments

The research is supported by the the National Key Basic Research Program of China under Grant No. 2013CB83440, the National Natural Science Foundation of China under Grants Nos. 11235001, 11320101004 and 11575007, the Open Project Program of State Key Laboratory of Theoretical Physics, Institute of Theoretical Physics, Chinese Academy of Sciences, China (No. Y4KF041CJ1), and the United Kingdom Science and Technology Facilities Council under grant No. ST/L005743/1.

References

[1] P. M. Walker and G. D. Dracoulis, Nature **399**, 35 (1999).
[2] P. M. Walker and G. D. Dracoulis, Hyperfine Interact. **135**, 83 (2001).
[3] F. R. Xu, P. M. Walker, J. A. Sheikh, and R. Wyss, Phys. Lett. B **435**, 257 (1998).
[4] M. W. Reed *et al.*, Phys. Rev. Lett. **105**, 172501 (2010).
[5] F. R. Xu, E. G. Zhao, R. Wyss, and P. M. Walker, Phys. Rev. Lett. **92**, 252501 (2004).
[6] H. L. Liu, F. R. Xu, S. W. Xu, R. Wyss, and P. M. Walker, Phys. Rev. C **76**, 034313 (2007).
[7] S. W. Xu *et al.*, Phys. Rev. C **71**, 054318 (2005).
[8] S. Aberg, Nucl. Phys. A **306**, 89 (1978).
[9] F. R. Xu, P. M. Walker, and R. Wyss, Phys. Rev. C **59**, 731 (1999).
[10] Y. Shi, F. R. Xu, H. L. Liu, and P. M. Walker, Phys. Rev. C **82**, 044314 (2010).
[11] A. N. Deacon *et al.*, Phys. Rev. C **76**, 054303 (2007).
[12] N. Alkhomashi *et al.*, Phys. Rev. C **80**, 064308 (2009).
[13] F. R. Xu, P. M. Walker, and R. Wyss, Phys. Rev. C **62**, 014301 (2000).
[14] H. L. Liu, F. R. Xu, P. M. Walker, and C. A. Bertulani, Phys. Rev. C **83**, 067303 (2011).
[15] J. Dudek, Z. Szymanski, T. Werner, Phys. Rev. C **23**, 920 (1981).
[16] P. Moller, J. R. Nix, Nucl. Phys. A **536**, 20 (1992).
[17] F. R. Xu, W. Satula, and R. Wyss, Nucl. Phys. A **669**, 119 (2000).
[18] F. R. Xu, Chin. Phys. Lett. **18**, 750 (2001).
[19] F. R. Xu, P. M. Walker, and R. Wyss, Phys. Rev. C **65**, 021303 (R) (2002).

[20] Y. H. Zhang *et al.*, Phys. Rev. C **68**, 054313 (2003).
[21] G. D. Dracoulis *et al.*, Phys. Rev. C **87**, 014326 (2013).
[22] G. D. Dracoulis *et al.*, Phys. Lett. B **720**, 330 (2013).
[23] Z. Podolyák *et al.*, Phys. Rev. C **66**, 011304 (2002).
[24] Z. Podolyák *et al.*, Phys. Lett. B **491**, 225 (2000).
[25] G. D. Dracoulis *et al.*, Phys. Lett. B **635**, 200 (2006).
[26] C. Wheldon *et al.*, Phys. Rev. C **59**, R2334 (1999).
[27] C. Wheldon *et al.*, Nucl. Phys. A **652**, 103 (1999) .
[28] C. S. Purry *et al.* Nucl. Phys. A **672**, 54 (2000).
[29] H. L. Liu, F. R. Xu, P. M. Walker, and C. A. Bertulani, Phys. Rev. C **83**, 011303 (R) (2011).
[30] H. L. Liu, F. R. Xu, and P. M. Walker, Phys. Rev. C **86**, 011301(R) (2012).
[31] H. L. Liu and F. R. Xu, Phys. Rev. C **87**, 067304 (2013).
[32] P. M. Walker and F. R. Xu, Phys. Lett. B **635**, 286 (2006).
[33] F. R. Xu, R. Wyss, and P. M. Walker, Phys. Rev. C **60**, 051301(R) (1999).
[34] Y. Shi, P. M. Walker, and F. R. Xu, Phys. Rev. C **85**, 027307 (2012).
[35] S. Hofmann *et al.*, Eur. Phys. J. A **10**, 5 (2001).
[36] R.-D. Herzberg *et al.*, Nature **442**, 896 (2006).
[37] H. B. Jeppesen *et al.*, Phys. Rev. C **79**, 031303 (2009).
[38] H. L. Liu, P. M. Walker, and F. R. Xu, Phys. Rev. C **89**, 044304 (2014).
[39] Yue Shi, F. R. Xu, P. M. Walker, and G. D. Dracoulis, Phys. Rev. C **85**, 064304 (2012).
[40] X. M. Fu *et al.*, Phys. Rev. C **89**, 054301 (2014).
[41] X. M. Fu *et al.*, Phys. Rev. C **87**, 044319 (2013).

EPJ Web of Conferences **123**, 01003 (2016) DOI: 10.1051/epjconf/201612301003
Heavy Ion Accelerator Symposium 2015

QCD and a new paradigm for nuclear structure

A. W. Thomas[1,a]

[1] *CSSM and ARC Centre of Excellence for Particle Physics at the Terascale, School of Chemistry and Physics, University of Adelaide, Adelaide SA 5005, Australia*

Abstract. We review the reasons why one might choose to seriously re-examine the traditional approach to nuclear theory where nucleons are treated as immutable. This examination leads us to argue that the modification of the structure of the nucleon when immersed in a nuclear medium is fundamental to how atomic nuclei are built. Consistent with this approach we suggest key experiments which should tell us unambiguously whether there is such a change in the structure of a bound nucleon. We also briefly report on extremely promising recent calculations of the structure of nuclei across the periodic table based upon this idea.

1 Introduction

Since the discovery of the neutron in the 1930s, the overwhelming majority of theoretical studies of nuclear structure have adopted the hypothesis that the protons and neutrons inside a nucleus are immutable objects whose internal structure never changes. These immutable objects interact through non-relativistic two- and three-body forces and the challenge is primarily to accurately solve the many-body problem. The phenomenological forces used include physics such as Yukawa's pion exchange and as a consequence the precise calculation of observables may require the inclusion of exchange current corrections.

Beginning with the famous one-boson-exchange potentials [1], it became clear that the dominant part of the intermediate range attraction between nucleons had a Lorentz scalar, isoscalar character, which was phenomenologically represented by the exchange of a σ meson. For decades this meson was viewed as an artifact involving an unphysical meson used purely for convenience. However, careful dispersion relation treatments of πN scattering in the past decade have shown that this state does indeed exist [2]. Confirmation of this Lorentz scalar, isoscalar character of the intermediate range attraction in the NN force also came from dispersion relation studies by groups in Paris [3], Stony Brook and elsewhere. Walecka and co-workers exploited the Lorentz scalar nature of the NN attraction and the Lorentz vector character of the short range repulsion to build a very successful, fully relativistic theory of nuclear matter [4] and later finite nuclei [5]. Here too the nucleons were immutable.

All this was very satisfactory but for one vexatious issue. At nuclear matter densities the typical mean scalar field strength felt by a bound nucleon in the Walecka model is of order 500 MeV. *This is a huge number.* As a consequence, the effective mass of the bound nucleon is only one half of its free mass.

At around the same time as Walecka and collaborators developed their model, the theory of the strong interaction underwent a revolution. Quantum Chromodynamics was developed as a local gauge field theory built on color. It became clear that, by analogy with Rutherford's work on the nucleus within the atom, the natural explanation of the discovery of scaling at SLAC in the late 60's [6] was that the nucleon too was primarily empty space containing point-like quarks.

From this more fundamental point of view the huge scalar field experienced by a bound nucleon is even more challenging. How can it be that the exchange of a scalar meson, which must couple to the confined quarks in the nucleon with such strength, can have no effect on the internal structure of the nucleon, which after all is far from point-like?

Considerations like these led Guichon [7] to propose a dramatically different approach to nuclear binding, the Quark Meson Coupling (QMC) model, where the effect of the mean scalar field generated by other nucleons is treated self-consistently in solving for the wave function of each confined quark. Taking the simplest form for the coupling of the σ and ω mesons to quarks confined in the MIT bag model [8, 9], means that in nuclear matter the vector field simply shifts the definition of the energy, while the scalar field modifies the Dirac wave function. *This difference in the effect of the two Lorentz components of the nuclear mean field is crucial, as their effects more or less cancel when it comes to the total energy but for the quark motion (or loosely speaking, wave function) the scalar field is not cancelled.*

A critical effect of the change in the quark wave function induced by an attractive scalar field is that the size of the lower Dirac component increases. In turn this reduces the value of $\int dV \bar{\psi}\psi$, which defines the overall strength

[a]Supported by the Australian Research Council through grants DP150103101 and CE110001104.

DOI: 10.1051/epjconf/201612301003

with which the scalar field couples to the nucleon. This process is completely analogous to the way an atom rearranges its internal structure to oppose an applied electric field. Thus the parameter calculated within any particular quark model which describes this is called the "scalar polarizability", d. The overall scalar coupling to the nucleon is written in the simplest approximation as

$$g_{\sigma N}(\sigma) = g_{\sigma N}(0) - \frac{d}{2}(g_{\sigma N}(0)\sigma)^2 . \qquad (1)$$

In the MIT bag model $d \approx 0.22R$, with R the bag radius.

This behaviour is very straightforward and appears in all relativistic quark models used so far. Nevertheless, in terms of nuclear structure it is profound. Whereas the repulsion felt by each nucleon grows linearly with density, the scalar attraction saturates as the density rises and one naturally finds saturation of nuclear matter. This mechanism is both new and extremely effective. As a result the mean scalar field felt by a nucleon at the saturation density of nuclear matter is just a few hundred MeV, much lower than that found in the Walecka model.

Philosophically, this approach is radically different from anything done before because the colourless clusters of quarks which occupy single particle levels in nuclear matter may have nucleon quantum numbers but their internal structure is modified. Almost immediately it was shown [10] that this change could account for the key features of the famous nuclear EMC effect, discovered in the early 80's.

Later the model was developed further by Guichon, Rodionov and Thomas [11] to correctly treat the effect of spurious centre of mass motion in the bag, which had resulted in anomalously small ωN couplings. In the same paper the model was also extended to finite nuclei, showing very naturally how one obtains realistic spin orbit forces. Finally, since the model is built at the quark level, using the same quark model, with the same quark-meson coupling constants, one can derive the properties of *any* bound hadron. For example, it shows very naturally why the spin-orbit force for the Λ hyperon is extremely small [12, 13]. For a complete overview of the phenomenological consequences of the QMC model we refer to the review by Saito *et al.* [14].

With the motivation for the QMC approach clearly established, one is naturally led to the following lines of investigation. First, given the success of the conventional approach to nuclear structure based upon non-relativistic two- and three-body forces, it is natural to ask how that is related to QMC. We address this in section 2. Second, one may also ask what evidence there is to support the at first sight radical idea that the clusters of quarks bound in shell model orbits actually have internal structure different from that of a free nucleon. This is addressed in Section 3, where we anticipate the results of a critical experiment performed at Jefferson Lab, which are expected to appear soon. Section 4 summarises this new approach to the structure of the atomic nucleus and looks to further consequences of it.

2 Nuclear structure: a new force of the Skyrme type

It is worthwhile to begin with some remarks on the application of effective field theory (EFT) to nuclear structure, since that also is often treated as containing all of the consequences of QCD [15]. Certainly the systematic application of chiral effective field theory to the NN and NNN forces and hence to nuclear structure has proven quite powerful. Such an approach is built upon the symmetries of QCD and is often considered to be equivalent to it. The problem is that the EFT approach needs some power counting scheme, which is a purely human construction. It also needs a set of hadronic degrees of freedom (dof) and that choice too is at the whim of the user. Finally, the EFT typically applied to nuclear problems is non-relativistic.

The usual choice of dof are nucleons and pions. If these are indeed the appropriate dof one is in luck. However, given the remarks in the Introduction, where we saw that on model independent grounds the intermediate range attraction between nucleons is a rather large Lorentz scalar, this is not so obvious. The attractive scalar and repulsive vector forces may cancel (in the central component of the nuclear force) to produce a relatively small amount of binding but the effect of those two components on the internal structure of a nucleon is completely different.

In an EFT the only way to include the effect of a change in the structure of a bound nucleon at the level of QCD is to include nucleon excited states amongst the dof. Typically this is limited to the Δ resonance, where we do know the relevant couplings quite well. However, given that the σ meson has quantum numbers 0^{++}, one may expect that the inclusion of excitations like the Roper resonance [16] may be relevant. Unfortunately, we have so little knowledge of that state that, at the present time, it would be very difficult to include it in an EFT framework in any reliable manner.

As a consequence, building an EFT of nuclei based upon nucleon and pion dof may not be as accurate an expression of QCD as it may appear at first sight.

An alternative approach to developing an EFT for nuclear structure is based on the density functional approach. There one starts with the QMC model itself and develops a density functional equivalent to it. From this one can use the machinery developed around the Skyrme forces [17], which have proven so successful in the study of both nuclear structure and reactions.

Indeed, using the density functional approach it has proven possible to develop a clear connection between the self-consistent treatment of in-medium hadron structure and the existence of many-body [18] or density dependent [19] effective forces. Dutra *et al.* [20] critically examined a variety of phenomenological Skyrme models of the effective density dependent nuclear force against the most up-to-date empirical constraints. Amongst the few percent of the Skyrme forces studied which satisfied all of these constraints, the Skyrme model SQMC700, was unique in that it was actually *derived from the QMC model* and hence incorporated the effects of the internal structure of the nucleon and its modification in-medium.

DOI: 10.1051/epjconf/201612301003

Very recently, Stone, Guichon, Reinhard and Thomas [21] carried out a systematic study of the properties of atomic nuclei across the whole periodic table using the new, effective, density-dependent *NN* force derived from the QMC model [19]. The study began by defining those combinations of the three fundamental couplings in the model (namely the σ, ω and ρ couplings to the up and down quarks) which reproduce the saturation density, binding energy per nucleon and symmetry energy of nuclear matter within the empirical uncertainties on these quantities. Then, a search was carried out for the set of three parameters satisfying this nuclear matter constraint which best described the ground-state properties of a selection of more than 100 nuclei across the entire periodic table.

The root mean-square deviation of the fit from the actual binding energy for this set of nuclei was just 0.35%. For the superheavy nuclei where the binding energies are known, the deviation was a mere 0.1%. This level of agreement with the empirical binding energies is remarkable, in that it is comparable with the very best phenomenological Skyrme forces which have typically 11 or more adjustable parameters.

Not only does this derived effective *NN* force satisfactorily describe binding energies but going beyond the nuclei used in the fit it accurately describes the evolution of quadrupole deformation across isotopic chains, including shell closures. It also proved capable of describing the observed shape co-existence of prolate, oblate and spherical shapes in the Zr region. Finally, it naturally gave a double quadrupole-octupole phase transition in the Ra-Th region.

These are remarkable successes given the extremely small number of parameters and this suggests that it would be worthwhile to apply this derived effective force across a variety of challenges in modern nuclear physics.

3 Experimental tests

Almost immediately after the creation of the QMC model it was applied [10] to the modification of the valence quark distribution in nuclei discovered by the European Muon Collaboration (EMC), known as the EMC effect [22]. That early work was based on the MIT bag model, for which the calculation of structure functions is possible within some approximations [23] but complicated. More recently, the generalization of the QMC model to the NJL model, suggested by Bentz and Thomas [24], has also been applied to the EMC effect with similar success [25]. The modification of the quark wave functions within the bound nucleons, because of the applied mean scalar field, naturally suppresses the valence distributions at large Bjorken x.

While this approach is the *only* quantitative model of nuclear structure which is able to describe the nuclear EMC effect, it is not yet universally accepted as the explanation for it. For example, it has recently been suggested that the entire EMC effect should be attributed to an as yet uncalculable modification of the nucleons involved in short-range correlations [26], while the rest of the nucleons apparently remain totally unchanged.

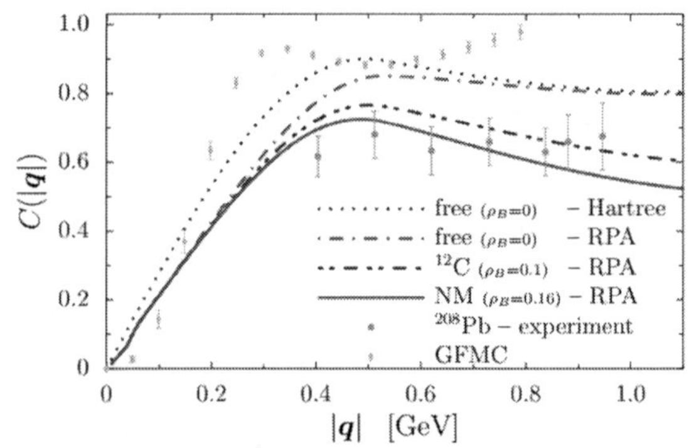

Figure 1. Predictions (from Ref. [36]) for the Coulomb sum rule as a function of three momentum transfer for nuclear matter at densities corresponding to ^{12}C and ^{208}Pb, with or without the effect of the in-medium modification of the nucleon electric form factors. Also shown are the GFMC calculations for ^{12}C (small points [37]) and older experimental data for ^{208}Pb [34, 38].

Another feature of this approach to nuclear structure is that the elastic form factors of the nucleon are also modified in-medium [27]. Using the QMC model, predictions were made almost 20 years ago for the experiment being planned at Jefferson Lab to measure the ratio of the electric to magnetic form factors of a proton bound in ^{4}He [28]. A decade later the measurements were in remarkably good agreement with those predictions [29–31], showing a significant medium modification. However, after the data appeared it was shown that it could also be fit by adding an unusually large polarised charge exchange correction. Although we are aware of no data supporting that proposed correction and no proposal to check it experimentally, it has muddied the waters sufficiently that this cannot yet be regarded as a "smoking gun".

Another suggestion, which seems far less susceptible to unknown nuclear corrections, involves the measurement of the longitudinal response function measured in inelastic electron scattering [34]. That was also examined in the late 90's on the basis of the modification of the electric form factor of the proton [32], already mentioned. Very recently, inspired by the proposal of Meziani and collaborators [33] to make a definitive measurement of this quantity for several nuclei across the periodic table, this response function and the associated Coulomb sum rule of McVoy and van Hove [35] were investigated using the NJL model to describe the structure of both the free and bound nucleons [36]. This work not only treated self-consistently the modification of bound nucleon structure resulting from the mean scalar field but it also included a state-of-the-art treatment of relativistic corrections and RPA correlations. The results are illustrated in Fig. 1.

At high values of the momentum transfer the effect of relativity and of the medium modification of the electric form factor of the proton in particular are both very significant. The older data certainly favours the new calculations and it is clearly vital to have the results of the comprehensive new experiment from Jefferson Lab as soon as

DOI: 10.1051/epjconf/201612301003

possible. The beauty of this particular measurement is that it appears to be extremely insensitive to other nuclear corrections, including the effect of short-range correlations.

4 Summary

We have presented a compelling argument that within the framework of QCD one is naturally led to the conclusion that the structure of a bound nucleon must differ from that in free space.

This idea has been used to derive, starting from the quark level, a new, density-dependent effective nuclear force which has proven remarkably accurate in describing the properties of finite nuclei across the entire periodic table, while at the same time reproducing the known properties of nuclear matter. We trust that these remarkable results will inspire a great deal more work on nuclear structure within this framework over the coming years.

We have seen that within the quantitative models of nuclear structure that have been developed within this approach, using either the MIT bag or the NJL model to describe nucleon structure, one finds a natural explanation of the nuclear EMC effect. There are also predictions for the modification of the electromagnetic form factors of the bound nucleon, for which the most unambiguous test is the Coulomb sum rule. There is an expectation that definitive new data for this will come from Jefferson Lab in the near future.

Finally, we briefly mention a number of other consequences of this approach to nuclear structure which are both fascinating and the subject of experimental investigation in the near future. For example, a careful study of nuclear structure functions has shown that this approach predicts an important isovector component of the nuclear EMC effect [39]. For a nucleus like ^{56}Fe this leads to a correction to the Paschos-Wolfenstein relation which is of the sign and magnitude to reduce the NuTeV anomaly by more than one standard deviation. These predictions will be tested directly in future measurements of parity violation [40] at Jefferson Lab following the 12 GeV upgrade.

Within this approach one also finds a remarkably large nuclear modification of the spin dependent parton distributions of the nucleon [41]. Again, future experiments planned at Jefferson Lab will test this through the measurement of the spin structure functions of light nuclei with an unpaired proton.

In conclusion, we stress that while one can derive effective *NN* forces which can be used in traditional nuclear structure calculations, *the underlying physics constitutes a new paradigm for nuclear theory*. The quark clusters which occupy shell model orbits in finite nuclei have internal structure which depends on the local scalar field – they are *not immutable*. This simple observation, which is entirely natural within the framework of QCD, explains the saturation of nuclear matter and the nuclear EMC effect and predicts a dramatic reduction in the Coulomb sum rule as well as a multitude of other phenomena which will be subject to experimental study in the coming decade.

Acknowledgements

I am indebted to the many collaborators who have contributed to the understanding of this approach to nuclear structure, particularly P. A. M. Guichon, W. Bentz, I. Cloët, K. Saito, J. Stone and K. Tsushima. This work was supported by the University of Adelaide and by the Australian Research Council through the ARC Centre of Excellence for Particle Physics at the Terascale (CE110001104), an ARC Australian Laureate Fellowship (FL0992247) and DP150103101.

References

[1] T. Ueda and A. E. S. Green, Phys. Rev. **174**, 1304 (1968).
[2] B. Ananthanarayan, G. Colangelo, J. Gasser and H. Leutwyler, Phys. Rept. **353**, 207 (2001).
[3] W. N. Cottingham, M. Lacombe, B. Loiseau, J. M. Richard and R. Vinh Mau, Phys. Rev. D **8**, 800 (1973).
[4] J. D. Walecka, Annals Phys. **83**, 491 (1974).
[5] B. D. Serot and J. D. Walecka, Phys. Lett. B **87**, 172 (1979).
[6] E. D. Bloom *et al.*, Phys. Rev. Lett. **23**, 930 (1969).
[7] P. A. M. Guichon, Phys. Lett. B **200**, 235 (1988).
[8] A. Chodos, R. L. Jaffe, K. Johnson, C. B. Thorn and V. F. Weisskopf, Phys. Rev. D **9**, 3471 (1974).
[9] A. W. Thomas, Adv. Nucl. Phys. **13**, 1 (1984).
[10] A. W. Thomas, A. Michels, A. W. Schreiber and P. A. M. Guichon, Phys. Lett. B **233**, 43 (1989).
[11] P. A. M. Guichon *et al.*, Nucl. Phys. A **601**, 349 (1996).
[12] K. Tsushima *et al.*, Nucl. Phys. A **630**, 691 (1998).
[13] K. Tsushima, K. Saito and A. W. Thomas, Phys. Lett. B **411**, 9 (1997) [Phys. Lett. B **421**, 413 (1998)].
[14] K. Saito, K. Tsushima and A. W. Thomas, Prog. Part. Nucl. Phys. **58**, 1 (2007).
[15] S. Weinberg, Nucl. Phys. B **363**, 3 (1991).
[16] D. Leinweber *et al.*, arXiv:1511.09146 [hep-lat].
[17] D. Vautherin and D. M. Brink, Phys. Rev. C **5**, 626 (1972).
[18] P. A. M. Guichon and A. W. Thomas, Phys. Rev. Lett. **93**, 132502 (2004).
[19] P. A. M. Guichon *et al.*, Nucl. Phys. A **772**, 1 (2006).
[20] M. Dutra *et al.*, Phys. Rev. C **85**, 035201 (2012).
[21] J. R. Stone, P. A. M. Guichon, P. G. Reinhard and A. W. Thomas, Phys. Rev. Lett. **116**, 092501 (2016).
[22] J. J. Aubert *et al.* [European Muon Collaboration], Phys. Lett. B **123**, 275 (1983).
[23] A. I. Signal and A. W. Thomas, Phys. Rev. D **40**, 2832 (1989).
[24] W. Bentz and A. W. Thomas, Nucl. Phys. A **696**, 138 (2001).
[25] I. C. Cloet, W. Bentz and A. W. Thomas, Phys. Lett. B **642**, 210 (2006).
[26] D. Wang *et al.* [PVDIS Collaboration], Nature **506**, 67 (2014).

DOI: 10.1051/epjconf/201612301003

[27] D. H. Lu, K. Tsushima, A. W. Thomas, A. G. Williams and K. Saito, Phys. Rev. C **60**, 068201 (1999).

[28] D. H. Lu, A. W. Thomas, K. Tsushima, A. G. Williams and K. Saito, Phys. Lett. B **417**, 217 (1998).

[29] S. Strauch *et al.* [Jefferson Lab E93-049 Collaboration], Phys. Rev. Lett. **91**, 052301 (2003).

[30] M. Paolone *et al.*, Phys. Rev. Lett. **105**, 072001 (2010).

[31] J. M. Udias and J. R. Vignote, Phys. Rev. C **62**, 034302 (2000).

[32] K. Saito, K. Tsushima and A. W. Thomas, Phys. Lett. B **465**, 27 (1999).

[33] Jefferson Lab experiment E05-110: http://hallaweb.jlab.org/experiment/E05-110/exp_home/

[34] J. Morgenstern and Z. E. Meziani, Phys. Lett. B **515**, 269 (2001).

[35] K. W. McVoy and L. Van Hove, Phys. Rev. **125**, 1034 (1962).

[36] I. C. Cloët, W. Bentz and A. W. Thomas, Phys. Rev. Lett. **116**, 032701 (2016).

[37] A. Lovato, S. Gandolfi, R. Butler, J. Carlson, E. Lusk, S. C. Pieper and R. Schiavilla, Phys. Rev. Lett. **111**, 092501 (2013).

[38] A. Zghiche *et al.*, Nucl. Phys. A **572**, 513 (1994) [Nucl. Phys. A **584**, 757 (1995)].

[39] I. C. Cloet, W. Bentz and A. W. Thomas, Phys. Rev. Lett. **102**, 252301 (2009).

[40] I. C. Cloet, W. Bentz and A. W. Thomas, Phys. Rev. Lett. **109**, 182301 (2012).

[41] I. C. Cloet, W. Bentz and A. W. Thomas, Phys. Rev. Lett. **95**, 052302 (2005).

EPJ Web of Conferences **123**, 01004 (2016)
Heavy Ion Accelerator Symposium 2015

DOI: 10.1051/epjconf/201612301004

Time-dependent Hartree-Fock Study of Octupole Vibrations in doubly magic nuclei

C. Simenel[1,a], J. Buete[1], and K. Vo-Phuoc[1]

[1] *Department of Nuclear Physics, Research School of Physics and Engineering, The Australian National University, Canberra, ACT 2601, Australia*

Abstract. Octupole vibrations are studied in some doubly magic nuclei using the time-dependent Hartree-Fock (TDHF) theory with a Skyrme energy density functional. Through the use of the linear response theory, the energies and transition amplitudes of the low-lying vibrational modes for each of the nuclei were determined. Energies were found to be close to experimental results. However, transition amplitudes, quantified by the deformation parameter β_3, are underestimated by TDHF. A comparison with single-particle excitations on the Hartree-Fock ground-state shows that the collective octupole vibrations have their energy lowered due to attractive RPA residual interaction.

1 Introduction

Nuclei can take all sorts of shapes: sphere, rugby ball, discus, pear, banana... They can also oscillate between these shapes, leading to nuclear vibrations. The description of these vibrations is a fundamental test of quantum many-body models. In addition, vibrations strongly affect nuclear reactions [1] as shown by coupled-channel calculations [2]. This has also been demonstrated with microscopic theory in the case of fusion [3–7] and fission [8–12]. For instance, couplings between relative motion and low-lying octupole states induce the formation of a neck, as illustrated in Fig. 1, which lowers the Coulomb barrier [7]. Reaction models should then incorporate a proper description of these vibrations in order to get a deep insight into the reaction mechanisms.

Our basic tool is the time-dependent Hartree-Fock (TDHF) mean-field theory proposed by Dirac [13]. The TDHF approach describes the time-dependent motion of each nucleon wave-function, assuming that each nucleon moves freely in a mean-field generated by all the other nucleons. First TDHF applications to study the dynamics of nuclear systems already showed the ability of the approach to describe collective vibrations [14, 15] (see also [16, 17] for reviews). Extensions including pairing correlations [18–22] as well as other components of the residual interaction [23–26] have also been used to describe nuclear vibrations in a dynamical way.

Several modern TDHF codes using realistic Skyrme energy density functionals are now available [27–30]. In order to study nuclear vibrations, these codes are used to compute the time evolution of various multipole moments after a small excitation of the nucleus. The linear response theory is then used to extract the energies and transition strengths of the first phonon of these vibrations. This approach allows for the study of both low-lying collective states and giant resonances, including their direct decay via nucleon emission [31, 32].

In the small amplitude (linear) regime, the TDHF theory is fully equivalent to the RPA [33] which is a standard tool to study collective vibrations. Nevertheless, TDHF can also be used to investigate non-linear effects beyond the RPA [34–36]. Although the approach is able to describe both low-lying and high-lying vibrations, applications have mostly focussed on giant resonances studies [15, 21, 22, 28–32, 37–41]. Despite their importance in understanding reaction mechanisms, only few applications have been made to investigate low-lying vibrations [18, 21].

In this contribution, we study low-lying octupole vibrations in doubly magic nuclei. Indeed, low-lying octupole modes are known to strongly couple to the relative motion and thus to affect the outcome of the reactions [42–45]. Note that low-lying quadrupole phonons are also known to affect fusion [1]. However, their specific study will be the subject of a future work (see also [21]).

2 TDHF simulations

Pairing correlations are absent in mean-field calculations of doubly magic nuclei due to large gaps at the Fermi level. The nucleus is initially in the Hartree-Fock (HF) ground-state computed with the EV8 code [46]. The interaction between the nucleons is determined from the SLy4 parametrisation of the Skyrme functional [47]. The dynamics is computed with the TDHF3D code [27] using the same functional. Details of the method can be found in Refs. [7, 17].

[a] e-mail: cedric.simenel@anu.edu.au

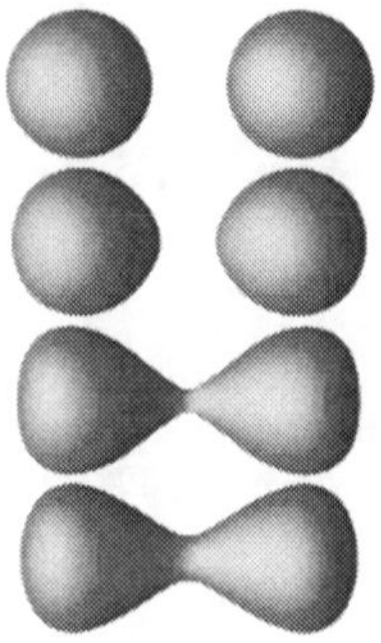

Figure 1. TDHF isodensity surfaces at half the saturation density, $\rho_0/2 = 0.08$ fm^{-3}, for a ^{40}Ca+^{40}Ca central collision at $E_{c.m.} = 53.3$ MeV. The time interval between each image is 0.25 zs.

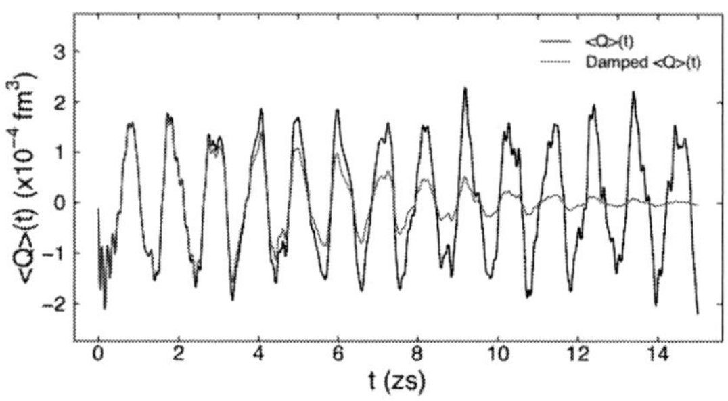

Figure 2. Time evolution of the octupole moment in ^{40}Ca following an octupole boost in the linear regime (black line). The red line shows the same moment multiplied by a damping function.

The vibrational modes are excited at the initial time with the operator $\varepsilon\hat{Q}$, where

$$\hat{Q} = \sqrt{\frac{7}{16\pi}}\sum_{i=1}^{A}[2\hat{z}^3 - 3\hat{z}(\hat{y}^2 + \hat{z}^2)] \quad (1)$$

is the octupole moment operator, A is the total number of nucleons and ε is the boost intensity. The latter is small enough to be in the linear regime, i.e., the induced moment $\langle\hat{Q}\rangle(t)$ is proportional to ε.

The calculations are performed on a three-dimensional grid with a plane of symmetry and a mesh spacing 0.8 fm. The calculations were run for 10,000 iterations with time step 1.5×10^{-24} s. As an example, the time evolution of the octupole moment in ^{40}Ca following an octupole boost at the initial time is shown in Fig. 2 (black line). A strong low-frequency oscillation is clearly observed.

The strength function is obtained from a Fourier transform of $\langle\hat{Q}\rangle(t)$. However, to avoid spurious oscillations due to the finite time of the TDHF calculation, the octupole moment is multiplied by a damping function so that the oscillations disappear at the final time (red line in Fig. 2). The resulting strength function is shown in Fig. 3. The main peak at ~ 3.9 MeV corresponds to the first 3^- in

Table 1. Energies of particle-hole excitations built on the ^{40}Ca HF ground-state and coupling to $J^\pi = 3^-$.

Configuration	Energy (MeV)
$\pi(1f_{\frac{7}{2}}, 1d_{\frac{3}{2}}^{-1})$	5.51
$\pi(1f_{\frac{7}{2}}, 2s_{\frac{1}{2}}^{-1})$	7.41
$\pi(1f_{\frac{7}{2}}, 1d_{\frac{5}{2}}^{-1})$	12.24
$\nu(1f_{\frac{7}{2}}, 1d_{\frac{3}{2}}^{-1})$	5.69
$\nu(1f_{\frac{7}{2}}, 2s_{\frac{1}{2}}^{-1})$	7.74
$\nu(1f_{\frac{7}{2}}, 1d_{\frac{5}{2}}^{-1})$	12.60

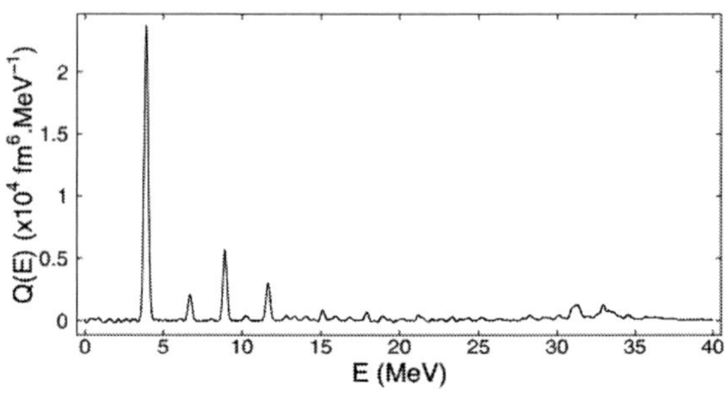

Figure 3. Strength function obtained by Fourier transform of the damped octupole moment in Fig. 2.

^{40}Ca. The latter is found experimentally at ~ 3.7 MeV [48], in good agreement with the calculations.

It is interesting to compare the energy of the first phonon of a vibration with the individual energies of the particle-hole excitations the phonon is built on, i.e., with a coupling to $J^\pi = 3^-$. The latter are reported in Table 1 for ^{40}Ca. All particle-hole excitation energies are significantly larger than the energy of the 3_1^- phonon. This shows that the RPA residual interaction is attractive for the corresponding octupole vibration.

In addition to the energy of the first phonon, the transition amplitude $q_3 = \langle 3_1^-|\hat{Q}|0_1^+\rangle$ between the 0^+ ground-state and the 3_1^- state can be obtained from the area of the corresponding peak in the strength function which is equal to $|q_3|^2$. Then, the value for the deformation parameter β_3 can be determined according to

$$\beta_3 = \frac{4\pi q_3}{3R_0^3 A}, \quad (2)$$

where R_0 is the nuclear radius. The latter is computed from the HF ground state using the isodensity at $\rho_0/2 = 0.08$ fm^{-3}. The deformation parameter of the 3_1^- state in ^{40}Ca is $\beta_3^{th.} \simeq 0.20$ from the present TDHF calculations. Despite large fluctuations in the experimental value $\beta_3^{exp.} \simeq 0.37 - 0.79$ which depends on the experimental probe [48], it remains that the TDHF prediction underestimates the data.

The energies and deformation parameters of the first octupole phonon are reported in Table 2 for some doubly-magic nuclei. Overall, the calculated energies are in

DOI: 10.1051/epjconf/201612301004

Table 2. Theoretical and experimental energy (in MeV) and deformation parameter of 3_1^- states in doubly-magic nuclei. Unless specified, the theoretical results are from this work. Experimental data are from [48] except for ^{132}Sn (ENSDF).

Nucleus	Ref.	$E_3^{th.}$	$\beta_3^{th.}$	$E_3^{exp.}$	$\beta_3^{exp.}$
^{16}O		7.23	0.30	6.13	0.37–0.79
^{40}Ca		3.92	0.20	3.74	0.30–0.41
^{48}Ca		5.74	0.11	4.51	0.25
^{56}Ni		9.59	0.13	(4.93)	-
^{100}Sn		6.91	0.08	-	-
^{132}Sn		5.53	0.07	(4.35)	-
^{208}Pb	[17]	3.4	0.05	2.61	0.11

Table 3. Same as Table 2 for ^{56}Ni.

Configuration	Energy (MeV)
$\pi(1g_{\frac{9}{2}}, 1f_{\frac{7}{2}}^{-1})$	11.761
$\pi(2p_{\frac{3}{2}}, 1d_{\frac{3}{2}}^{-1})$	11.768
$\pi(1f_{\frac{5}{2}}, 1d_{\frac{3}{2}}^{-1})$	15.080
$\nu(1g_{\frac{9}{2}}, 1f_{\frac{7}{2}}^{-1})$	12.181
$\nu(2p_{\frac{3}{2}}, 1d_{\frac{3}{2}}^{-1})$	12.262
$\nu(1f_{\frac{5}{2}}, 1d_{\frac{3}{2}}^{-1})$	15.710

reasonable agreement with experimental data, except for ^{56}Ni. However the experimental assignment to a 3^- state is uncertain. The origin of the large value of the 3^- energy in ^{56}Ni is due to the complete filling of the $1f_{7/2}$ proton and neutron shells. Then, the only possibility to produce a 3^- is to promote a $1f_{7/2}$ nucleon to the high-lying $1g_{9/2}$ shell, or to excite a nucleon from below the $Z, N = 20$ magic gap to the $1d$ shells. Both types of excitations require a large amount of energy, as shown in Table. 3.

Finally, the underestimation of the deformation parameter, already noted in ^{40}Ca, seems systematic. It would be interesting to investigate the origin of this effect which could come from a limitation of the mean-field approximation, from the choice of the interaction, or from the way this parameter is deduced experimentally.

3 Conclusions

Octupole vibrations in doubly-magic nuclei have been studied with TDHF and the linear response theory. The energy of the first phonons are in relatively good agreement with experimental data. In particular, variations induced by the underlying single-particle shell structure are well captured.

The deformation parameters, however, are underestimated in the present calculations. The origin of this discrepancy is unclear and deserves further studies. In particular, other parametrisations of the Skyrme functional should be tested.

Nevertheless, the TDHF results capture reasonably well the dependence of the phonon characteristics (energy and deformation parameter) with the mass of the nucleus. This indicates that TDHF is indeed a good tool in order to investigate the interplay between nuclear structure and reaction mechanisms. For instance, it was shown in Ref. [7] that, unlike ^{56}Ni, ^{40}Ca nuclei could develop octupole shapes in collision with a partner thanks to their low-lying collective 3^- state (see Fig. 1).

Acknowledgements

This work has been supported by the Australian Research Council under Grant No. FT120100760. This research was undertaken on the NCI National Facility in Canberra, Australia, which is supported by the Australian Commonwealth Government.

References

[1] M. Dasgupta, D. J. Hinde, N. Rowley, A. M. Stefanini, Ann. Rev. Nucl. Part. Sci. **48**, 401 (1998).
[2] K. Hagino, N. Takigawa, Prog. Th. Phys. **128**, 1061 (2012).
[3] C. Simenel, P. Chomaz, G. de France, Phys. Rev. Lett. **86**, 2971 (2001).
[4] C. Simenel, P. Chomaz, G. de France, Phys. Rev. C **76**, 024609 (2007).
[5] V. E. Oberacker, A. S. Umar, J. A. Maruhn, P. G. Reinhard, Phys. Rev. C **85**, 034609 (2012).
[6] C. Simenel, R. Keser, A. S. Umar, V. E. Oberacker, Phys. Rev. C **88**, 024617 (2013).
[7] C. Simenel, M. Dasgupta, D. J. Hinde, E. Williams, Phys. Rev. C **88**, 064604 (2013).
[8] C. Simenel, A. S. Umar, Phys. Rev. C **89**, 031601 (2014).
[9] G. Scamps, C. Simenel, D. Lacroix, Phys. Rev. C **92**, 011602 (2015).
[10] P. Goddard, P. Stevenson, A. Rios, Phys. Rev. C **92**, 054610 (2015).
[11] Y. Tanimura, D. Lacroix, G. Scamps, Phys. Rev. C **92**, 034601 (2015).
[12] A. S. Umar, V. E. Oberacker, C. Simenel, Phys. Rev. C **92**, 024621 (2015).
[13] P. A. M. Dirac, Proc. Camb. Phil. Soc. **26**, 376 (1930).
[14] P. Bonche, S. Koonin, J. W. Negele, Phys. Rev. C **13**, 1226 (1976).
[15] J. Błocki, H. Flocard, Phys. Lett. B **85**, 163 (1979).
[16] J. W. Negele, Rev. Mod. Phys. **54**, 913 (1982).
[17] C. Simenel, Eur. Phys. J. A **48**, 152 (2012).
[18] B. Avez, C. Simenel, P. Chomaz, Phys. Rev. C **78**, 044318 (2008).
[19] S. Ebata, T. Nakatsukasa, T. Inakura, K. Yoshida, Y. Hashimoto, K. Yabana, Phys. Rev. C **82**, 034306 (2010).
[20] I. Stetcu, A. Bulgac, P. Magierski, K. J. Roche, Phys. Rev. C **84**, 051309 (2011).
[21] G. Scamps, D. Lacroix, Phys. Rev. C **88**, 044310 (2013).
[22] G. Scamps, D. Lacroix, Phys. Rev. C **89**, 034314 (2014).
[23] M. Tohyama, Prog. Theor. Phys. **94**, 147 (1995).

[24] D. Lacroix, S. Ayik, P. Chomaz, Phys. Rev. C **63**, 064305 (2001).
[25] M. Tohyama, A. S. Umar, Phys. Lett. B **549**, 72 (2002).
[26] D. Lacroix, S. Ayik, P. Chomaz, Prog. Part. Nucl. Phys. **52**, 497 (2004).
[27] K. H. Kim, T. Otsuka, P. Bonche, J. Phys. G **23**, 1267 (1997).
[28] A. S. Umar, V. E. Oberacker, Phys. Rev. C **71**, 034314 (2005).
[29] T. Nakatsukasa, K. Yabana, Phys. Rev. C **71**, 024301 (2005).
[30] J. A. Maruhn, P. G. Reinhard, P. D. Stevenson, J. R. Stone, M. R. Strayer, Phys. Rev. C **71**, 064328 (2005).
[31] P. Chomaz, N. V. Giai, S. Stringari, Phys. Lett. B **189**, 375 (1987).
[32] B. Avez, C. Simenel, Eur. Phys. J. A **49**, 76 (2013).
[33] P. Ring, P. Schuck, *The Nuclear Many-Body Problem* (Springer Verlag, 1980).
[34] C. Simenel, P. Chomaz, Phys. Rev. C **68**, 024302 (2003).
[35] P. G. Reinhard, L. Guo, J. A. Maruhn, Eur. Phys. J. A **32**, 19 (2007).
[36] C. Simenel, P. Chomaz, Phys. Rev. C **80**, 064309 (2009).
[37] D. Lacroix, A. Mai, P. von Neumann-Cosel, A. Richter, J. Wambach, Phys. Lett. B **479**, 15 (2000).
[38] P. D. Stevenson, M. R. Strayer, J. Rikovska Stone, W. G. Newton, Int. J. Mod. Phys. E **13**, 181 (2004).
[39] D. Almehed, P. D. Stevenson, J. Phys. G **31**, S1819 (2005).
[40] P. G. Reinhard, P. D. Stevenson, D. Almehed, J. A. Maruhn, M. R. Strayer, Phys. Rev. E **73**, 036709 (2006).
[41] P. D. Stevenson, D. Almehed, P. G. Reinhard, J. A. Maruhn, Nucl. Phys. A **788**, 343 (2007).
[42] C. R. Morton, M. Dasgupta, D. J. Hinde, J. R. Leigh, R. C. Lemmon, J. P. Lestone, J. C. Mein, J. O. Newton, H. Timmers, N. Rowley *et al.*, Phys. Rev. Lett. **72**, 4074 (1994).
[43] J. R. Leigh, M. Dasgupta, D. J. Hinde, J. C. Mein, C. R. Morton, R. C. Lemmon, J. P. Lestone, J. O. Newton, H. Timmers, J. X. Wei *et al.*, Phys. Rev. C **52**, 3151 (1995).
[44] K. Hagino, N. Takigawa, M. Dasgupta, D.J. Hinde, J.R. Leigh, Phys. Rev. Lett. **79**, 2014 (1997).
[45] N. Rowley, K. Hagino, Nucl. Phys. A **834**, 110c (2010), the 10th International Conference on Nucleus-Nucleus Collisions (NN2009).
[46] P. Bonche, H. Flocard, P. H. Heenen, Comp. Phys. Com. **171**, 49 (2005).
[47] E. Chabanat, P. Bonche, P. Haensel, J. Meyer, R. Schaeffer, Nucl. Phys. A **635**, 231 (1998).
[48] T. Kibédi, R. H. Spear, At. Dat. Nucl. Dat. Tab. **80**, 35 (2002).

EPJ Web of Conferences **123**, 01005 (2016) DOI: 10.1051/epjconf/201612301005
Heavy Ion Accelerator Symposium 2015

Probes of shape transitions from mass and charge radii of nuclear ground states

B.H. Sun[1,a] and C.Y. Liu[1]

[1] *School of Physics and Nuclear Energy Engineering, Beihang University, Beijing 100191, China*

Abstract. The masses and sizes of nuclear ground states constitute two of the most precise and extensive arrays of experimental information. These data make a model-independent view of microscopic nuclear structure possible. Relevant differential observables of nuclear mass and charge radius can be highly sensitive to nuclear shape transitions. In this contribution, we examine the correlation of these two bulk properties to nuclear shape transitions. By combining different observables, it is even possible to isolate shape transitions from nuclear shell closures.

1 Introduction

One of the goals of nuclear physics is to understand how structure evolves in atomic nuclei with changing numbers of their proton and neutron constituents. This investigation dates back to the early days of the field. It was fundamental, for example, to the development of the shell model with the identification of magic numbers, and also to the earliest studies of collective structure. Today it remains one of the backbones for nuclear structure studies, especially in rare nuclei, far from the valley of stability. Because of its major role in the classification of nuclei, the unexpected onset of nuclear deformation and shape coexistence are very important. Experimentally, searching for unambiguous evidence of shape transitions or coexistence is thus crucial to reveal the single-particle level and structure changes [1, 2].

Both the mass and size of an atomic nucleus are bulk properties that reflect the total effect of all nucleons inside and their interactions [3–5]. Rearrangements of nucleons in nuclei due to, for example, the onset of deformation, or changes in single-particle levels, may manifest as subtle changes in these ground-state properties. Moreover, it is possible to magnify the subtle changes by using selected differential properties of nuclear masses and sizes. Just as in nuclear spectroscopic studies that use excited level structure, the measurement of nuclear masses and sizes to ultra-high precision can be a sensitive probe of sudden changes in the nuclear surface. Correlations of different observables in both ground states and excited states have been found (see e.g. [6]).

Considering that both the quality and quantity of experimental data on nuclear ground-state properties have increased tremendously over the last decade, it becomes increasingly useful to evaluate the systematic correlation between mass, size and shape transitions. In many cases, these two properties have been studied only recently for rare isotopes. In this work, we limit our investigation to nuclear ground-state properties, in particular the nuclear charge radius, which reflects in particular the distribution of protons inside an atomic nuclide. For this purpose, we take several isotopic chains at $Z \sim 40$, 62 and 82 as examples to illustrate the different probes of shape transitions. In this work, we refer to shape transitions quite generally as effects induced by changes in nuclear shapes, e.g., shape coexistence. For the sake of completeness, we will also show spectroscopic observables for comparison.

This paper is organized as follows. In Section 2, we first show the finite-difference formula that will be used to extract the effect of nuclear shape transition. Then in Section 3, we examine the signatures of shape coexistence from both mass and charge radius for three different zones in nuclear chart. Finally we give a summary in Section 4.

2 Probes for nuclear shape variances

We first discuss the key observables that will be used to identify shape transitions. Unless otherwise indicated, the data are taken from the the most recent completions of atomic mass and charge radius evaluations [7, 8]. Energies of the first-excited 2^+ states $[E(2_1^+)]$ and the transition probabilities between the first 2^+ and 0^+ ground states $[B(E2)]$ are from Ref. [9].

In principle, except where there is shell closure, phase transition, or neutron-proton pairing, or where the Wigner effect is present, mass or charge radii are a smooth function of Z and N. Therefore, finite-differences of masses and charge radii are a convenient means to identify subtle changes induced by variances in nuclear structure.

Two-neutron separation energies $S_{2n}(Z, N)$ are frequently used to avoid oscillations between odd and even

[a] e-mail: bhsun@buaa.edu.cn

nuclei due to pairing. Other finite-difference formula of nuclear masses used in this work include the effective neutron shell gap $G_{2n}(Z,N)$ and the empirical interaction strength between last two valence neutrons and two protons $\delta V_{2p-2n}(Z,N)$. These two quantities can be understood in the framework of the independent particle shell-model. They are defined as:

$$\begin{aligned} G_{2n}(Z,N) =& S_{2n}(Z,N) - S_{2n}(Z,N+2) \approx 4\frac{\partial M^2(Z,N)}{\partial N^2}, \\ \delta V_{2p-2n}(Z,N) =& -\frac{1}{4}\,[M(Z,N) + M(Z-2,N-2) \\ & -M(Z,N-2) - M(Z-2,N)] \\ \approx& -\frac{1}{4}\frac{\partial M^2(Z,N)}{\partial N \partial Z}, \end{aligned} \quad (1)$$

where $M(Z,N)$ is the mass of nuclide (Z,N). δV_{2p-2n} can be generalized to δV_{ip-jn} between the last i proton(s) and the last j neutron(s) [10–14].

As for nuclear charge radii $R_{ch}(Z,N)$, similar difference formula exist. The kink strength $S(Z,N)$ is the second differential of the mean-square charge radius:

$$S(N) = [R(N+2)-2R(N)+R(N-2)]A^{2/3} \approx 4\frac{\partial^2 R(N)}{\partial N^2}A^{2/3}, \quad (2)$$

where the $A^{2/3}$ term accounts for the mass dependence. Very recently, we proposed a new set of differential relations δR_{ip-jn} based on neighboring nuclei [15]:

$$\begin{aligned} \delta R_{ip-jn}(Z,N) &= R(Z,N) + R(Z-i,N-j) \\ &\quad -R(Z,N-j) - R(Z-i,N) \\ &\propto \frac{\partial R^2(Z,N)}{\partial N \partial Z}. \end{aligned} \quad (3)$$

Calculation of $\delta R_{ip-jn}(Z,N)$ for a single nuclide (Z,N) will be done in three different forms, i.e., $\delta R_{1p-1n}(Z,N)$, $\delta R_{1p-2n}(Z,N)$, and $\delta R_{2p-1n}(Z,N)$. When several predictions exist for a specific nuclide, the mean value $\overline{\delta R}_{p-n}(Z,N)$ is then calculated and treated afterwards for this nuclide.

3 Selected examples at (Z ~ 40, N ~ 60), (Z ~ 64, N ~ 90) and (Z ~ 82)

Here we show the ground state and excited states observables as function of neutron number, namely, two-neutron separation energies, charge radii, δV_{2p-2n}, $\overline{\delta R}_{p-n}$, $E(2^+_1)$, and $B(E2)$, for three selected zones characterized by nuclear shape changes. Because there are similarities in the following discussions of selected zones, it is useful first to summarize the general features of different observables at occurrences of changes in nuclear structure.

S_{2n} decreases smoothly with neutron number between two neutron magic numbers, and deviations from this behavior is a manifestation of changes in microscopic nuclear structure. For example, a sharp decrease of S_{2n} (i.e., decreased binding of the last two neutrons), most probably indicates a neutron shell closure or mid-shell region. A subtle change in the slope of S_{2n} or a flattening in S_{2n} normally indicate collective effects. As a second derivative of nuclear masses, G_{2n} is useful to wipe out the smooth components. It is often used to show the strength of shell effect.

Similar to nuclear masses, nuclear charge radii are almost linearly dependent on neutron number between two neutron magic numbers, while a local minimum is found at magic numbers. Characteristic slope changes (kinks) of charge radii, especially the rapid increase of charge radii, have been found to be correlated to deformation effect. This is particularly striking in the Hg and Pt nuclei as will shown below.

δV_{2p-2n} and $\overline{\delta R}_{p-n}$ involve at least four neighboring nuclei along two isotopic chains. They are thus useful to examine the local evolution of structure. More specifically, δV_{2p-2n} values can help to explain and describe shell structure and orbit occupations near the Fermi surface, while deviations of $\overline{\delta R}_{p-n}$ from zero are found to be correlated to onset of deformation or shape coexistence. As we will show below, these different probes can be correlated and complimentary to each other.

Spectroscopic fingerprints — excited-state energies and γ-ray transition strengths — have played a leading role in the exploration of shape transitions, as they present patterns characteristic of the underlying nuclear shape. Here we choose the two most important observables: $E(2^+_1)$ and $B(E2)$. However, data on $E(2^+_1)$ and $B(E2)$ are much more sparse in comparison with the ground state properties as will be seen below.

The region of $Z \sim 40$ and $N \sim 60$ is known to be dominated by the sudden onset of deformation around $N = 60$. Figure 1 shows the different observables discussed above. On addition of neutrons to the magic shell gap at $N = 50$, nuclei undergo transitions from spherical to prolate (and oblate) deformations as $N = 60$ is approached and crossed. The onset of deformation is clearly seen by a kink in S_{2n}, and a rapid increase in nuclear size R_{ch}. This effect is more visible in G_{2n} and S as a significant deviation from $G_{2n} \approx 1$ and $S_z \approx 0$. The impact of the onset of deformation in δV_{2p-2n} is similar to the shell closure at $N = 50$, i.e., a significantly drop in magnitude from more than 0.6 MeV to about 0.4 MeV. This is understandable because the empirical proton-neutron interaction is sensitive to the underlying nuclear levels, especially the valence nucleons, and both the shell closure and the rapid change in deformation can lead to similar changes in occupied single-particle orbits.

$\overline{\delta R}_{p-n}$ is particularly interesting. Very different from all the other observables, it is not sensitive to closed shells at all. This deviation might be due to the interplay of neutrons and protons, which somehow washes out the shell effects in charge radii. On the other hand, $\overline{\delta R}_{p-n}$ is apparently correlated with the onset of deformation, where there are large deviations from zero. Combined analysis of $\overline{\delta R}_{p-n}$ and the other observables like S, can even isolate the effects caused by closed shell closure and phase transitions.

The nuclides in the region of $Z \sim 64$ and $N \sim 90$ are centered on the stability line and thus are one of the most intensive and varied spectroscopic and theoretical

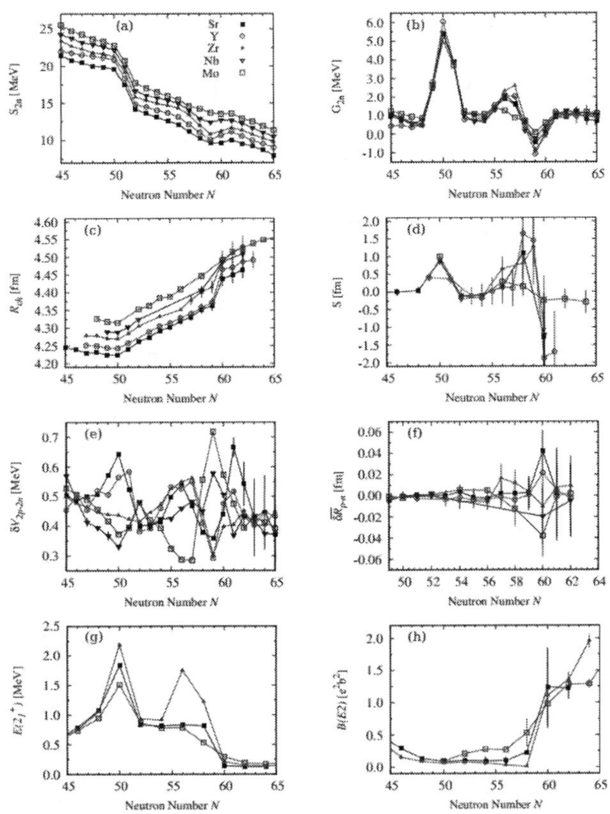

Figure 1. (a) Two-neutron separation energies (S_{2n}), (b) neutron shell gap (G_{2n}), (c) charge radii (R_{ch}), (d) kink strength (S), (e) effective interaction strength between last two valence neutrons and two valence protons ($\delta V_{2p\text{-}2n}$), (f) a new charge radius relation ($\overline{\delta R}_{p-n}$), (g) energy of the first-excited 2^+ state ($E(2_1^+)$), and (h) transition probability between the first 2^+ and the 0^+ ground state ($B(E2)$), for selected isotopic chains across the $N \sim 60$ region.

studies of collectivity. Figure 2 shows the systematics of all observables for this region. All observables exhibit a near identical pattern to that observed in the range of ($Z \sim 40, N \sim 60$) except $\delta V_{2p\text{-}2n}$. The nuclear-mass region around $Z = 82$ is characterized by the most extensive manifestation of shape coexistence. This effect can be immediately seen from extreme increases and "staggering" in charge radii and the relevant derivatives at the $N \sim 104$ region in Fig. 3. The sensitivity of $\overline{\delta R}_{p-n}$ to nuclear phase transitions is also shown. Correlations to shape coexistence are also shown in S_{2n}, G_{2n} and $\delta V_{2p\text{-}2n}$, but it seems that R_{ch} is much more sensitive to shape coexistence than S_{2n}.

An interesting point from Figs. 1-3 is that the simple relation of $\overline{\delta R}_{p-n} = 0$ holds to good precision except at occurrences of shape transitions. In other words, such relations offer a very useful feature in prediction of the root-mean-square (rms) charge radius of a given nuclide in terms of known charge radii of its neighboring nuclei. As an example, in Fig. 4 we show the difference between the predictions of charge radius using $\overline{\delta R}_{p-n} = 0$ and experimental values. The predictions of $\overline{\delta R}_{p-n} = 0$ are remarkably good throughout the nuclear chart, especially for nuclides heavier than Ni. However, larger deviations are seen at, e.g., ($Z \sim 40, N \sim 60$), ($Z \sim 55, N \sim 40$), ($Z \sim 64, N \sim 90$), and $Z \sim 80$, which are correlated with shape transitions. Moreover, the charge radii of 325 nuclides can be estimated as indicated in Fig. 4 within an accuracy of typically 0.01 fm. However, when sudden variances occur in nuclear shape, or where such changes are expected, one must be cautious about the predicitons of unknown charge radii. In these cases, it may be instructive to check the correlations with other ground and excited state observables.

4 Summary

To summarize, we examine possible hints of shape transition in ground state observables, i.e., masses and charge radii. In particular, we show that the recent proposed $\overline{\delta R}_{p-n}$ can be a good indicator for sudden changes in nuclear shape, and seems not to be affected by large shell closures. It is interesting to extend the analysis of $\overline{\delta R}_{p-n}$ in the near future over a wider range of isotopes, including exotic ones. Combined analysis of these different probes from mass and charge radii is always valuable, as these

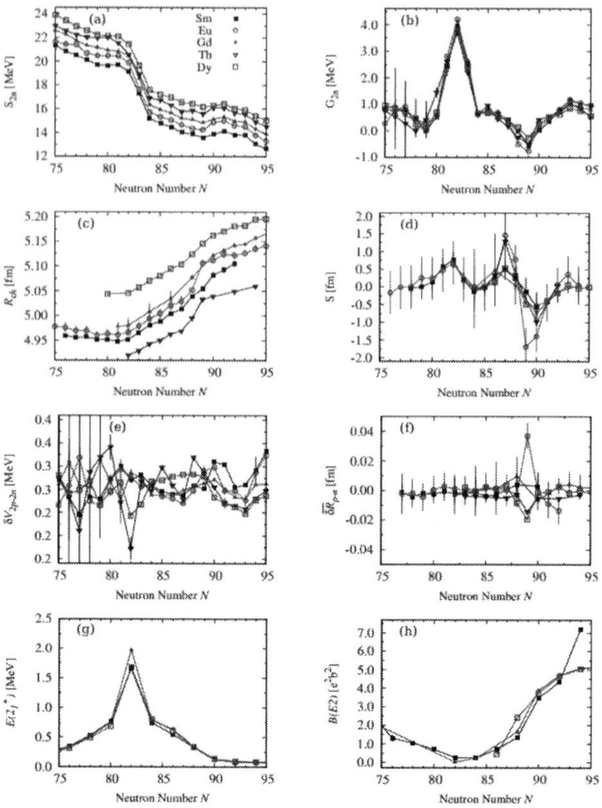

Figure 2. Same as Fig. 1, but for selected isotopic chains across the $N \sim 90$ region. To guide the eye, the large uncertainties of charge radii in Tb and Dy isotopic chains are not shown.

properties are often the first obtained for exotic nuclei in regions far from stability.

Acknowledgements

This work was supported partially by the National Natural Science Foundation of China (No. 11475014) and the Fundamental Research Funds for the Central Universities.

References

[1] P. Cejnar, J. Jolie, R. F. Casten, Rev. Mod. Phys. **82**, 2155 (2010).

[2] K. Heyde and J. L. Wood, Rev. Mod. Phys. **83**, 1467 (2011).

[3] I. Angeli *et al.*, J. Phys. G **36**, 085102 (2009).

[4] K. Blaum, Phys. Rep. **425**, 1 (2006).

[5] B.H. Sun, Yu. A. Litvinov, I. Tanihata, Y. H. Zhang, Front. Phys. **10**, 102102 (2015).

[6] R. B. Cakirli, R. F. Casten, K. Blaum, Phys. Rev. C **82**, 061306(R) (2010).

[7] M. Wang, G. Audi, A. H. Wapstra, F. G. Kondev, M. MacCormick, X. Xu, and B. Pfeiffer, Chin. Phys. C **36**, 1603 (2012).

[8] I. Angeli and K. P. Marinova, At. Data Nucl. Data Tables **99**, 69 (2013).

[9] B. Pritychenko, M. Birch, B. Singh, M. Horoi, At. Data Nucl. Data Tables **107**, 1 (2016).

[10] H. Jiang, G. J. Fu, Y. M. Zhao, and A. Arima, Phys. Rev. C **82**, 054317 (2010).

[11] G. J. Fu, Y. Lei, H. Jiang *et al.*, Phys. Rev. C **84**, 034311 (2011).

[12] H. Jiang, G. J. Fu, B. Sun *et al.*, Phys. Rev. C **85**, 054303 (2012).

[13] G. J. Fu, J. J. Shen, Y. M. Zhao, and A. Arima, Phys. Rev. C **87**, 044309 (2013).

[14] Y. Y. Cheng, Y. M. Zhao, and A. Arima, Phys. Rev. C **91**, 024314 (2015).

[15] B. H. Sun, Y. Lu, J. P. Peng, C. Y. Liu, and Y. M. Zhao, Phys. Rev. C **90**, 054318 (2014); ibid, Phys. Rev. C **91**, 019902(E) (2015).

DOI: 10.1051/epjconf/201612301005

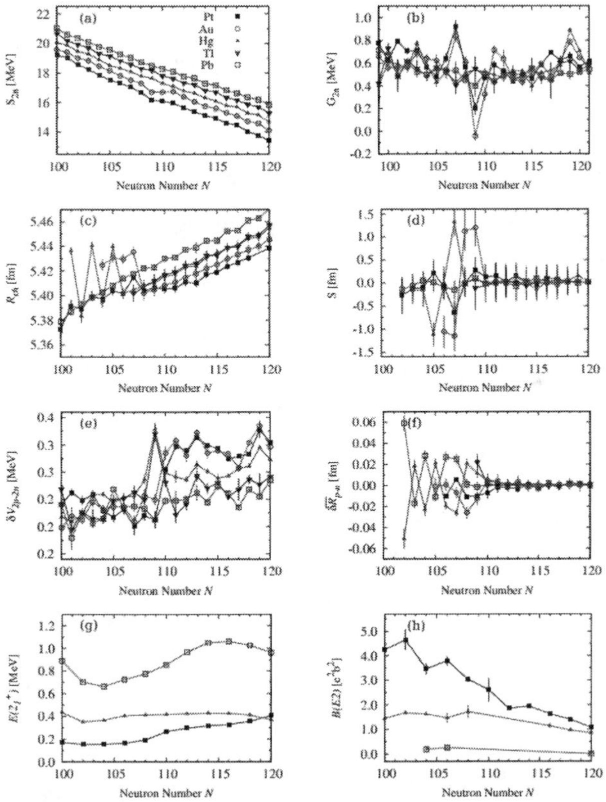

Figure 3. Same as Fig. 1, but for selected isotopic chains across the $Z \sim 82$ region. To guide the eye, the large uncertainties of $\overline{\delta R}_{p-n}$ charge radii in the $^{184-187}$Pt isotopic chain are not shown.

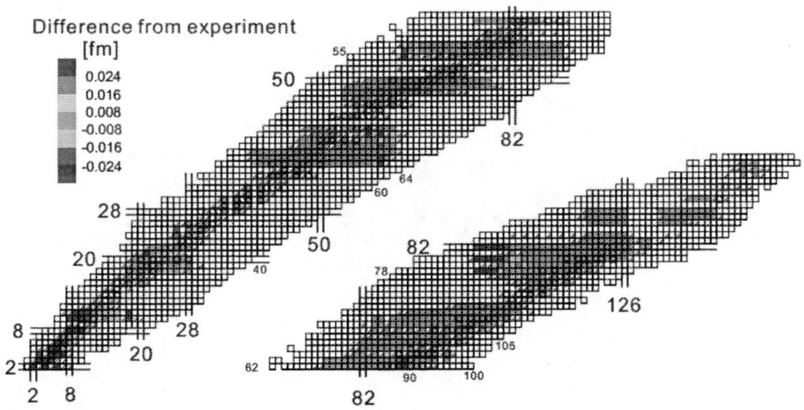

Figure 4. Chart of the nuclides showing the differences (colour coded, in fm) between the predictions of charge radii using $\overline{\delta R_{p-n}} = 0$ and experimental values. The squares outlined in black indicate stable isotopes. The magic numbers are also indicated. 325 nuclides whose rms charge radius can be predicted within an accuracy around 0.01 fm are marked by the green triangles. From Ref. [15].

EPJ Web of Conferences **123**, 01006 (2016) DOI: 10.1051/epjconf/201612301006
Heavy Ion Accelerator Symposium 2015

Evolution of collective structures in the heavy transitional nuclei above the N = 82 closed shell

D. T. Joss[1,a]

[1] *Oliver Lodge Laboratory, Department of Physics, University of Liverpool, Liverpool, L69 7ZE, United Kingdom*

Abstract. The emergence of collectivity in atomic nuclei is a fundamental paradigm in nuclear structure physics. Considerable progress has been made towards the identification of excited states in heavy neutron-deficient nuclei above the $N = 82$ closed shell. This paper summarises recent progress in the spectroscopic study of the Ta, W and Re nuclides with $N \sim 90$ obtained from recoil and recoil-decay tagging experiments. The nuclei near $N = 90$ occupy a γ-soft transitional region where the nuclear shape is particularly sensitive to the interplay between collective excitations and the underlying single-particle structure. The consequences of these interactions for low and high-spin states are discussed.

1 Introduction

The identification of excited states in atomic nuclei spanning complete shells is crucial for understanding the evolution of nuclear structure from both experimental and theoretical perspectives. The character of the low-lying excited states changes as the number of valence nucleons increases beyond closed shells from single-particle multiplets to collective excitations arising from the correlated motion of many nucleons. The $82 \leq N \leq 126$ shell presents the largest range of nuclei where the properties of excited states can be measured experimentally with current techniques. In particular, the advent of selective experimental methods has allowed the identification of excited states at the extremes of this shell where the onset or demise of collective behaviour occurs [1, 2].

Much progress has been made towards identifying the excited states in the neutron-deficient nuclei above the $N = 82$ shell gap. However, extending knowledge to those isotopic chains heavier than Hf is challenging due to their relatively low production cross sections in fusion evaporation reactions. For example, excited states have been established in the closed shell nuclide ^{154}Hf ($N = 82$) [3] but the lightest isotopes with low-lying excited states in heavier even-Z chains are ^{160}W [4], ^{162}Os ($N = 86$) [5], and ^{168}Pt ($N = 90$) [6]. Moreover, collective effects and the underlying single-particle structure cannot be completely decoupled and the identification of excited states in odd-A nuclei and their spectroscopic study, particularly in the transitional region, can provide more detailed insights into nuclear shapes and correlations. This paper summarises recent results for the even mass W [7] and odd mass Re [8] and Ta [9, 10] nuclides in the $N \sim 90$ transitional region obtained from recoil and recoil-decay tagging experiments and discusses the implications for nuclear structure in this region.

[a]e-mail: david.joss@liverpool.ac.uk

2 Experimental Methodology

Excited states in several neutron-deficient Ta, W and Re nuclides were populated using fusion-evaporation reactions in experiments performed at the Accelerator Laboratory of the University of Jyväskylä, Finland. Prompt γ rays were detected at the target position by the Jurogam γ-ray spectrometer [11]. The recoiling fusion-evaporation residues were separated from fission products and scattered beam by the RITU gas-filled recoil separator [12] and implanted into the double-sided silicon strip detectors (DSSD) of the GREAT spectrometer [13] at the focal plane. Recoiling nuclei were distinguished from the residual scattered beam and radioactive decays by energy loss and (in conjunction with the DSSDs) time of flight methods using the GREAT multiwire proportional counter.

All detector signals from Jurogam and GREAT were passed to the total data readout acquisition system [14] where they were time stamped with a precision of 10 ns to allow accurate temporal correlations between γ rays detected at the target position, recoil implants at the focal plane and their subsequent radioactive decays. These triggerless data were sorted into γ-γ matrices and γ-γ-γ cubes using GRAIN [15] and analysed with the RADWARE software packages [16].

The recoil-decay tagging (RDT) technique correlates γ rays detected at the target position with the characteristic radioactive decays of the residual nucleus at the focal plane of a recoil separator [1]. The RDT technique provides high-confidence correlations under the optimal conditions of short decay half-lives and high decay branching ratios. Gamma-ray emissions from ^{162}W, ^{163}Re and ^{165}Re

DOI: 10.1051/epjconf/201612301006

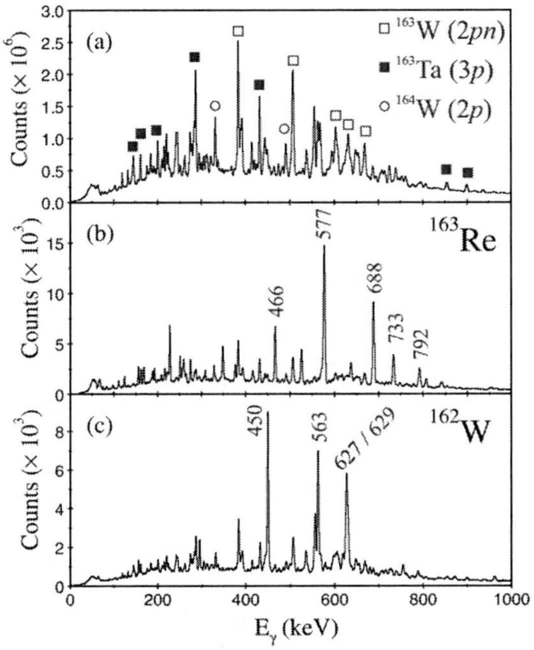

Figure 1. Gamma rays detected in the Jurogam spectrometer following the ^{106}Cd(^{60}Ni, xp,yn,$z\alpha$) reaction at 270 MeV. (a) γ rays in delayed coincidence with any recoil implantation detected in the GREAT DSSDs located at the RITU focal plane. Gamma rays are associated with the main reaction products in the legend. (b) Gamma rays correlated with recoil implantations followed by the characteristic α decay from the $11/2^-$ isomer in ^{163}Re within the same DSSD pixel of the GREAT spectrometer. The recoil-α correlation time was limited to 600 ms. (c) Gamma rays correlated with recoil implantations followed by the distinct ground-state α decay of ^{162}W within the same DSSD pixel of the GREAT spectrometer. The recoil-α correlation time was limited to 2.4 s.

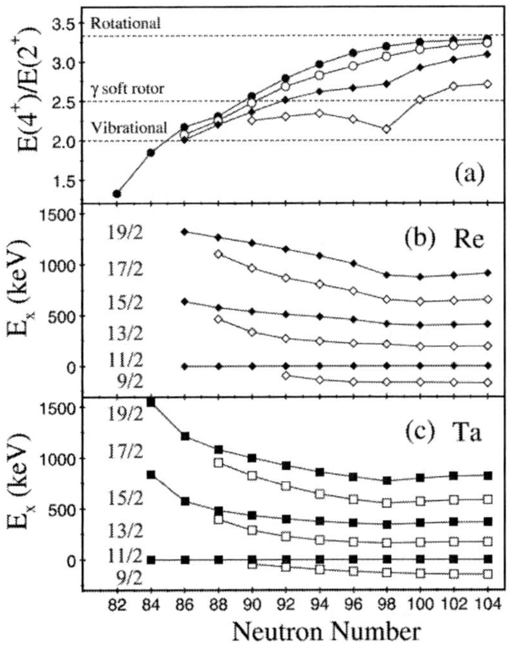

Figure 2. (a) Ratios of the 4^+ and 2^+ state excitation energies as a function of neutron number for the Hf (filled circles), W (open circles), Os (filled diamonds) and Pt (open diamonds) nuclides. The limiting cases for vibrational, γ-soft and rotational nuclei are indicated by dashed lines. Excitation energies of the low-lying excited states in the $\pi h_{11/2}$ bands of the neutron-deficient (b) Re isotopes and (c) Ta isotopes. The level energies are plotted relative to the $11/2^-$ state. Open (filled) symbols correspond to states in the $\alpha = +1/2$ ($\alpha = -1/2$) signature partner.

could be identified by correlations with their characteristic α decays. Other nuclides with less favourable α-decay properties, such as ^{161}Ta, ^{163}Ta and ^{164}W were studied using γ rays detected in Jurogam that were in delayed coincidence with recoils implanted in the focal plane detector. Figure 1 shows examples of spectra obtained with the RDT technique.

3 The Heavy N > 82 Transitional Nuclei

3.1 Low-Lying Excited States

The development of collective behaviour outside the N = 82 shell gap is reflected in the variation of low-lying excited states as a function of neutron number. The ratio of the 4^+ and 2^+ state excitation energies in even-Z, even-mass nuclei, $E(4^+)/E(2^+)$, is often used as an indicator of collectivity. Figure 2(a) shows the variation of the $E(4^+)/E(2^+)$ ratio across the Hf, W, Os and Pt isotopes against the reference limits for vibrational, γ-soft and rotational nuclei. Isotones with higher atomic numbers have lower $E(4^+)/E(2^+)$ ratios as a consequence of their smaller number of valence protons. The Hf, W and Os isotopes show a smooth increase in $E(4^+)/E(2^+)$ as a function of neutron number indicating the emergence of collectivity from single-particle type excitations (N ~ 82), through γ-soft triaxial shapes (N ~ 90) to axially symmetric deformed rotors near the neutron midshell at (N = 104). The N ~ 98 platinum isotopes (and their osmium isotones to a lesser extent) show a marked deviation from the gradual increase in $E(4^+)/E(2^+)$ ratio with increasing neutron number. This arises from perturbations to the yrast sequences arising from mixing between coexisting bands [17–19].

The transitional nuclei near N = 90 are soft to triaxial deformation due to spatial density distributions of the proton and neutron orbitals at the top or bottom of their respective shells [20]. For example, in $^{164}W_{90}$, the proton Fermi surface lies in a region of low level density near the high-Ω $h_{11/2}$ states while the neutron Fermi surface lies close to the lowest-Ω $\nu i_{13/2}$ orbital and negative-parity orbitals originating from both the $\nu f_{7/2}$ and $\nu h_{9/2}$ subshells. Further evidence for soft triaxial shapes at N ~ 90 can be found by probing the structure of the neighbouring odd-mass Re and Ta nuclides. Figure 2(b) and 2(c) show the variation of the excitation energies of the low-lying yrast states in the Re and Ta isotopes, respectively.

The lightest even-N Re and Ta isotopes have ground states that are near-spherical and based on the low-Ω $s_{1/2}$ or $d_{3/2}$ proton states with a low-lying isomer formed by exciting the odd proton into the high-Ω $h_{11/2}$ state [21–26]. Gamma rays associated with the characteristic radioactive decays of the $h_{11/2}$ isomer in ^{157}Ta [27] ($N = 84$), ^{159}Ta [4] and ^{161}Re [28] ($N = 86$) have revealed non-collective excited states formed by coupling the odd proton to the few valence neutrons in $f_{7/2}$, $h_{9/2}$ and $i_{13/2}$ states. In these nuclei the odd-proton is decoupled from the underlying W cores. With the addition of a few neutrons, coupled bands with a large degree of signature splitting are observed in the $88 \leq N \leq 94$ nuclides [8–10, 29–31]. This splitting is attributed to triaxiality, which relieves the degeneracy of the high-Ω $h_{11/2}$ orbital. As the maximum number of valence neutrons is reached at $N = 104$, the $\alpha = +1/2$ ($13/2^-$ and $15/2^-$) states lie near the midpoints of the levels in the $\alpha = -1/2$ signature partner [32, 33]. This is interpreted as a restoration of signature degeneracy in the $\pi h_{11/2}$ orbital due to the onset of axial prolate symmetry arising from maximal correlations between nucleons at the neutron midshell.

3.2 High-Spin Structures

The yrast cascades for the transitional ($N = 88, 90$) Ta and W nuclides are observed to spins beyond the first rotational alignment. At present, excited states in their Re isotones have not been observed to sufficiently high spin to characterise rotationally aligned structures. The high-spin structure of the $N \sim 90$ nuclides can be discussed in terms of the γ-ray energy, E_γ, as a function of the angular momentum of the initial emitting state I_i, see Fig. 3. This formulation is chosen over the familiar alignment plots as a function of rotational frequency since it relies entirely on experimental data and removes the need to specify K, which is ill-defined in triaxial nuclei.

A comparison of Fig. 3(a) and 3(b) shows that band 1 in ^{162}W has a lower alignment gain than band 1 in its heavier even-mass neighbour, ^{164}W. Gamma-ray spectroscopy studies of $N < 90$ nuclei suggest that the low average deformation of these nuclei favours excitations involving the negative-parity ($\nu f_{7/2}$, $\nu h_{9/2}$) states [10, 34–38]. The quadrupole deformation of ^{164}W is predicted be larger ($\beta_2 = 0.161$) than ^{162}W ($\beta_2 = 0.116$) [39], which is sufficient to bring the lowest Ω $\nu i_{13/2}$ orbital close to the Fermi surface. Indeed, the $i_{13/2}$ neutron orbital dominates the yrast spectra of the $N \geq 90$ W isotopes where excitations of one or two $\nu i_{13/2}$ quasineutrons are favoured at low spin and excitation energy [7, 36, 40, 41].

The excited band (band 2) in ^{162}W has a similar degree of alignment to the excited bands (bands 2 and 3) in ^{164}W. The excited bands in the W isotopes are interpreted as two quasiparticle configurations formed by coupling a single $\nu i_{13/2}$ orbital to the available $\nu f_{7/2}, \nu h_{9/2}$ negative-parity states. The excited bands have angular momentum values that are greater than the ($\nu f_{7/2}$, $\nu h_{9/2}$)^2 aligned configuration of ^{162}W yet lower than the ($i_{13/2}$)^2 aligned configuration in ^{164}W.

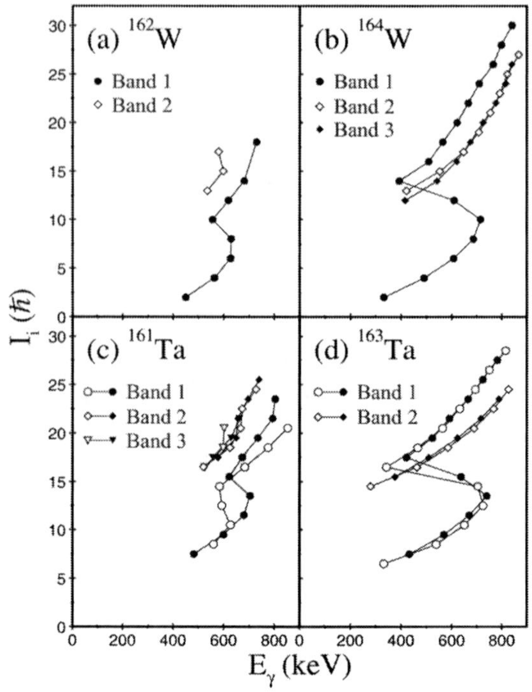

Figure 3. Angular momentum of the initial emitting state I_i versus γ-ray energy E_γ for collective bands in (a) ^{162}W, (b) ^{164}W, (c) ^{161}Ta, (d) ^{163}Ta. Open (filled) symbols correspond to states in the $\alpha = +1$ ($\alpha = -1$) signature partner in the even-Z, even-A W isotopes or the $\alpha = +1/2$ ($\alpha = -1/2$) signature partners in the odd-Z, odd-A Ta isotopes

Figures 3(c) and 3(d) show that similar features are observed in the Ta isotones resulting from coupling an single $h_{11/2}$ quasiproton to the underlying W core configurations. The excited bands in ^{161}Ta (bands 2 and 3) show increases in angular momenta that are consistent with the initial stages of a transition from three to five quasiparticle configurations. It is apparent from Fig. 3(c) that there is considerable signature splitting both below and above the first rotational alignment in ^{161}Ta. Signature splitting at low spin is also observed in ^{163}Ta, which vanishes at high spin following the ($i_{13/2}$)^2 alignment. This indicates that the rotational alignment of $\nu i_{13/2}$ neutrons has a greater polarising effect on the soft-triaxial core than the ($\nu f_{7/2}$, $\nu h_{9/2}$) negative-parity alignments.

4 Summary

This paper has reviewed the recent progress made in understanding the development of collectivity in the transitional Ta, W and Re isotopes located above the $N = 82$ closed neutron shell. Level schemes in these nuclei have been established and extended in a campaign of recoil (-decay) tagging experiments at the University of Jyväskylä

Accelerator Laboratory using the Jurogam and GREAT spectrometers in conjunction with the RITU gas-filled separator.

The evolution of low-spin states as a function of neutron number reveals the onset of collectivity from single-particle excitations to well-deformed collective rotors. The nuclei in the transitional ($N \sim 90$) region between these two regimes display features that are consistent with γ-soft triaxial rotors.

The high-spin structure of the transitional Ta and W isotopes highlights the interplay between collective excitations and specific nucleon orbitals at the Fermi surface. A comparison of the rotational properties of $^{162}W_{88}$ and $^{164}W_{90}$ indicates that the $\nu f_{7/2}, h_{9/2}$ rotational alignment is favoured over the $\nu i_{13/2}$ pair alignment at $N = 88$. An increase in deformation at $N = 90$ alters the relative positions of the $\nu f_{7/2}, h_{9/2}$ and $\nu i_{13/2}$ orbitals and configurations involving the latter become favoured for the $N \geq 90$ W isotopes. Spectroscopic investigations of the odd-Z nuclides $^{161}Ta_{88}$ and $^{163}Ta_{90}$ highlight the sensitivity of the γ-soft core to the occupation of core-polarizing orbitals at the Fermi surface. The low-spin states in both nuclei are single $h_{11/2}$ quasiproton configurations that exhibit a significant degree of signature splitting. It is shown that while the rotational alignment of a pair of $f_{7/2}, h_{9/2}$ neutrons in ^{161}Ta does little to alter the γ-soft triaxial shape, the $\nu i_{13/2}$ pair alignment in ^{163}Ta is sufficient to polarise the triaxial core towards an axially symmetric prolate shape.

Acknowledgments

The author would like to acknowledge G. D. Dracoulis, G. J. Lane, J. Burde, M. A. Deleplanque, A. O. Macchiavelli, J. Oliviera and F. S. Stephens for kindly sharing information related to their work on ^{162}W from their HERA experiments at the Lawrence Berkeley National Laboratory. The author would like to express his gratitude to the staff of the Accelerator Laboratory at the University of Jyväskylä for their excellent technical support. Emily Higgins is acknowledged for providing the spectra used in Figure 1. Financial support for this work has been provided by the UK Science and Technology Facilities Council (STFC), the EU 6th Framework Programme, Integrating Infrastructure Initiative - Transnational Access, (EURONS, Contract No. 506065), the Academy of Finland under the Finnish Centre of Excellence Programme, and by the Swedish Research Council. The author is indebted to the UK/France (STFC/IN2P3) Loan Pool and the GAMMAPOOL European Spectroscopy Resource for the loan of detectors for Jurogam.

References

[1] E. S. Paul *et al.*, Phys. Rev C **51**, 78 (1995).
[2] S. J. Steer *et al.*, Phys. Rev C **84**, 044313 (2011).
[3] J. H. McNeill *et al.*, Phys. Rev. Lett. **63**, 860 (1989).
[4] A. Keenan *et al.*, Phys. Rev. C **63**, 064309 (2001).
[5] D. T. Joss *et al.*, Phys. Rev. C **70**, 017302 (2004).
[6] M. B. Gomez Hornillos *et al.*, Phys. Rev. C **79**, 064314 (2009).
[7] D. T. Joss *et al.*, Phys. Rev. C **93**, 4307 (2016).
[8] T. R. Davis-Merry *et al.*, Phys. Rev. C **91**, 034319 (2015).
[9] M. Sandzelius *et al.*, Phys. Rev. C **80**, 054316 (2009).
[10] K. Lagergren *et al.*, Phys. Rev. C **83**, 014313 (2011).
[11] C. W. Beausang and J. Simpson, J. Phys.G **22**, 527 (1996).
[12] M. Leino *et al.*, Nucl. Instrum. and Meth. in Phys. Res. B **99**, 653 (1995).
[13] R. D. Page *et al.*, Nucl. Instrum. and Meth. in Phys. Res. B **204**, 634 (2003).
[14] I. H. Lazarus *et al.*, IEEE Transactions on Nuclear Science **48** 567 (2001).
[15] P. Rahkila, Nucl. Instrum. and Meth. in Phys. Res. **A595**,637 (2008).
[16] D. C. Radford, Nucl. Instrum. and Meth. in Phys. Res. **A361**, 297 (1995).
[17] G. D. Dracoulis *et al.*, Phys. Rev. C **44**, R1246 (1991).
[18] G. D. Dracoulis *et al.*, J. Phys. G: Nucl. Phys **12**, L97 (1986).
[19] G. D. Dracoulis *et al.*, Phys. Lett. B **115**, 367 (1982).
[20] Y. S. Chen, S. Frauendorf, and G. A. Leander, Phys. Rev. C **28**, 2437 (1983).
[21] D. T. Joss *et al.*, Phys. Lett. B **641**, 34 (2006).
[22] R. D. Page *et al.*, Phys. Rev. Lett. **68**, 1287 (1992).
[23] I. G. Darby *et al.*, Phys. Rev. C **83**, 064320 (2011).
[24] R. J. Irvine *et al.*, Phys. Rev. C **55**, R1621 (1997).
[25] R. D. Page *et al.*, Phys. Rev. C **53**, 660 (1996).
[26] A. Thornthwaite *et al.*, Phys. Rev. C **86**, 064315 (2012).
[27] D. Seweryniak *et al.*, Phys. Rev. C **71**, 054319 (2005).
[28] K. Lagergren *et al.*, Phys. Rev. C **74**, 024316 (2006).
[29] D. T. Joss *et al.*, Phys. Rev. C **68**, 014303 (2003).
[30] X. H. Zhou *et al.*, Eur. Phys. J. A **19**, 11 (2004).
[31] D. G. Roux *et al.*, Phys. Rev. C **63**, 024303 (2001).
[32] D. E. Archer *et al.*, Phys. Rev. C **52**, 1326 (1995).
[33] J. R. Leigh *et al.*, Nucl. Phys. Rev. **A183**, 177 (1972).
[34] G. D. Dracoulis *et al.*, *Proceedings of the International Conference of Nuclear Structure at High Angular Momentum*, (Ottawa, 1992), and *AECL Report No. 10613* (unpublished, 1992), Vol. 2, p. 94.
[35] H. J. Li *et al.*, Phys. Rev. C **92**, 014326 (2015).
[36] J. Thomson *et al.*, Phys. Rev. C **81**, 014307 (2010).
[37] J. M. Rees *et al.*, Phys. Rev. C **83**, 044314 (2011).
[38] M. C. Drummond *et al.*, Phys. Rev. C **87**, 054309 (2013).
[39] P. Möller, J.R. Nix, W.D. Myers, and W.J. Swiatecki, At. Data Nucl. Data Tables **59**, 297 (1995).
[40] J. Simpson *et al.*, J. Phys. G **18**, 1207 (1992).
[41] K. Theine *et al.*, Nucl. Phys. **A548**, 71 (1992).

Proton emission – new results and future prospects

R. D. Page[1,a]

[1] *Department of Physics, Oliver Lodge Laboratory, University of Liverpool, Liverpool L69 7ZE, United Kingdom*

Abstract. Proton emission is the radioactive decay mode that is expected to determine the limit of observable proton-rich nuclei for most elements. Considerable progress has been made in the study of proton-emitting nuclei since the first observation of direct proton emission nearly 50 years ago. This has led to improvements in our understanding of this decay process and provided invaluable nuclear structure data far from the valley of beta stability. The rapid fall in half-lives with increasing neutron deficiency when proton emission dominates makes it likely that for some elements, the lightest isotopes whose ground states can be observed in conventional experiments have already been reached. The enhanced stability against proton emission of the recently discovered high-lying isomer in ^{158}Ta raises the possibility that proton emission from multiparticle isomers could be observed in nuclei beyond the expected boundaries of the nuclear landscape.

1 Introduction

Isomers are a widespread feature across the chart of nuclides and their study can provide an important testing ground for a range of nuclear theories [1]. The existence of isomeric states is also an important consideration in the study of nuclei beyond the proton drip line and is inextricably linked to the question of the limits of observable nuclei. The first experimentally observed case of direct proton emission was from a high-spin isomer in ^{53}Co [2, 3]. In this unique case, the proton emitter's ground state is bound to proton emission, but the excitation energy of the proton-unbound isomer is sufficient to overcome the retarding effect of the centrifugal barrier and results in a proton-decay branching ratio of ~1.5 %. Subsequent studies of proton emission have primarily focused on low-lying states in nuclei beyond the proton drip line. As these studies reach the limits of nuclei that are accessible using recoil separators, attention may again switch to proton emission from high-lying isomeric states, in nuclei far beyond the proton drip line.

2 Proton emission from low-lying states

The discovery of the first cases of proton emission from nuclear ground states [4, 5] followed a decade after the ^{53}Co results and many further proton emitters have been discovered since [6]. Approximately half of the presently known cases of direct one-proton emission have been found in the region beyond the proton drip line that is bounded by the $N = 82$ and $Z = 82$ shell closures. In total, 21 cases of direct proton emission in this region have been reported to date. One reason for this rich sample is that either the proton-decaying states can be populated through alpha decays or the daughter nuclides are alpha emitters. This makes searches for proton-decay peaks with low numbers of counts very sensitive, because the alpha decays have characteristic energies that can be cleanly selected for correlation analyses and they tend to have fairly short half-lives, which result in relatively low levels of background from false correlations.

The structure of nuclei in this region is governed at low excitation energies by valence neutrons in the $\nu f_{7/2}$ and $\nu h_{9/2}$ orbitals and protons in the $\pi s_{1/2}$, $\pi d_{3/2}$ and $\pi h_{11/2}$ orbitals. In odd-Z nuclides, two states are generally found at low excitation energies that can decay by proton, alpha or beta emission (see Fig. 1a). In odd-A nuclei, the states usually have either an unpaired $s_{1/2}$ or $h_{11/2}$ proton, while in odd-odd nuclei either a $d_{3/2}$ or $h_{11/2}$ proton is coupled with an $f_{7/2}$ neutron to form the low-lying states. According to the Nordheim rules [7, 8], the lowest-energy state formed by coupling a $d_{3/2}$ proton with an $f_{7/2}$ neutron would have spin and parity 2^-, while the lowest-energy state when coupling an $h_{11/2}$ proton with an $f_{7/2}$ neutron would be 9^+. The energy differences between the states have been found to be $\lesssim$300 keV (see Fig. 1b), so direct gamma decays from the isomeric state to the ground state in both odd-A and odd-odd nuclides would be very slow owing to the large spin difference. Consequently charged particle emission generally dominates from the low-lying isomeric states and approximately half of the known cases of proton emission in this region are from these isomers, with the remainder being from ground states.

This pattern of two low-lying states is generally borne out, as can be seen in Fig. 1a, except for a few of the most exotic nuclei where only the $\pi h_{11/2}$ state has been observed. This is thought to be because the low-spin ground states are too short-lived to survive the flight time through a recoil separator [12–15, 17, 18]. The

[a]e-mail: rdp@ns.ph.liv.ac.uk

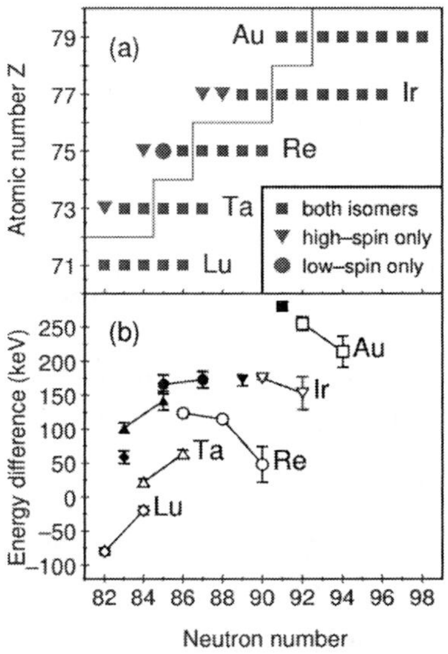

Figure 1. (a) Chart indicating the observed low-lying isomers in odd-*Z* nuclei. The triangles indicate nuclei in which only the high-spin isomer has been observed, the squares indicate those in which both the high-spin and low-spin isomers have been observed and the circle indicates ^{160}Re, which is the only nuclide in this region where only the low-spin isomer has been observed. Nuclides to the left of the line are known proton emitters. (b) Energy differences between the $\pi h_{11/2}$ isomers and the low-spin ($\pi s_{1/2}$ or $\pi d_{3/2}$) isomers. The low-spin isomers represent the ground states of the nuclides shown, except for the Lu isotopes for which the $\pi h_{11/2}$ configuration forms the ground state. The open symbols denote odd-*A* nuclei, while odd-odd nuclei are indicated by the filled symbols. Data are taken from Refs. [9–16].

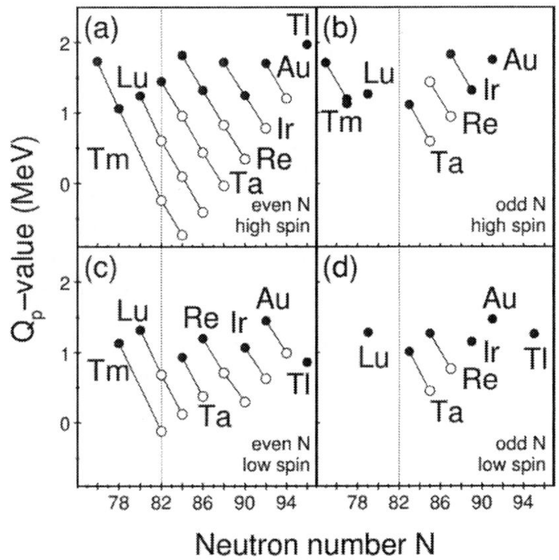

Figure 2. Experimental proton-decay Q-values of odd-*Z* elements from Tm to Tl measured directly are shown by the filled circles. Values estimated by combining these values with α-decay measurements are indicated by the empty circles. (a) is for high-spin ($\pi h_{11/2}$) states in odd-*Z*, even-*N* nuclei, while (b) shows the corresponding values for odd-*Z*, odd-*N* nuclei. (c) and (d) show the values for low-spin ($\pi s_{1/2}$ or $\pi d_{3/2}$) states in odd-*Z*, even-*N* and odd-*Z*, odd-*N* nuclei, respectively. Data are taken from Refs. [4, 5, 9–18, 21, 23–32].

only other exception is the case of ^{160}Re, for which proton and alpha-particle emission have only been observed from the $d_{3/2}$ ground state [19–21]. In this nuclide a 2.8-microsecond gamma-decaying isomer was identified [20] and subsequent in-beam gamma-ray spectroscopy measurements suggested that the isomer was based on a $\pi h_{11/2}$ orbital [22]. The discovery of the alpha-decay branch from the $\pi h_{11/2}$ state in ^{164}Ir established the excitation energy of the corresponding state in ^{160}Re, which was interpreted as being the isomeric state [15]. It seems that in ^{160}Re the multiplets of states formed by coupling the odd neutron and proton are so arranged as to allow a gamma-ray decay path that has a much shorter half-life than either proton or alpha emission from the $\pi h_{11/2}$ isomeric state [20].

Measurements of proton-decay energies can be used to deduce Q_p values that can be compared with predictions of atomic mass models. The systematics of Q_p values determined from proton-decay measurements on elements from Tm to Tl are shown in Fig. 2. For Ta, Re, Ir, Au and Tl proton emitters, the prevalence of extensive alpha-decay chains above $N = 82$ allows Q_p values to be determined indirectly for states that do not themselves have a known proton-decay branch. This is in contrast to the proton-emitting Tm and Lu nuclides whose Q_p values cannot be used in this way to deduce values for other nuclides. The smooth behaviour of the Q_p values as a function of neutron number allows local extrapolations of the trends to be made, which can be used to predict the decay properties of as yet unobserved states [33]. This makes it possible to identify good candidates for new proton emitter searches and to evaluate whether the limits of observable nuclei using conventional recoil separator techniques may have been reached.

3 Proton emission from high-lying isomers

In addition to the low-lying isomeric states discussed in the previous section, higher-lying multiparticle isomers are also known in this region. For example, seniority 3 $[\pi h^3_{11/2}]27/2^-$ isomers have been identified in the $N = 82$ isotones ^{149}Ho, ^{151}Tm and ^{153}Lu [34–36]. The possibility of proton emission from the corresponding isomer in ^{155}Ta to states in ^{154}Hf is discussed in Ref. [18]. The proton decay directly to the ground state of ^{154}Hf would be favoured by having the largest Q_p value, but be hindered by the large centrifugal barrier associated with the spin difference between the initial and final states. Decays to excited, higher-spin states in ^{154}Hf would be less hindered because the centrifugal barrier would be lower, but the excitation energy would reduce the effective Q_p value. In general, the balance between these competing effects will determine which proton-decay paths are most favoured in the decay of a given isomer.

Multiparticle isomers that decay by alpha emission have been established in $N = 84$ isotones [16, 37, 38]. In even-even nuclei these isomers are interpreted as having a $[\nu f_{7/2} \nu h_{9/2}]8^+$ structure, while in the odd-Z isotones the $\pi h_{11/2}$ proton couples with this configuration to give an isomeric $25/2^-$ state. Both ^{155}Lu and ^{157}Ta are unbound to proton emission (see Fig. 2) and the probability of proton-decay branches to states in their respective daughter nuclides are analysed in Ref. [39]. In the case of ^{155}Lu, proton emission from the isomer would be faster than from lower-lying states, which are only just unbound, whereas for ^{157}Ta all proton-decay paths from the isomer have longer partial half-lives than low-lying states.

The alpha and gamma decays of an even more complicated isomeric state in ^{158}Ta have been reported recently [39, 40]. A possible structure for this isomer is $[\pi h^{-3}_{11/2} \nu f_{7/2} \nu h_{9/2} \nu i_{13/2}]19^-$, analogous to that proposed for an isomer in ^{152}Ho [41]. This isomer in ^{158}Ta lies at an excitation energy of 2668 keV and although it is unbound to proton emission by 3261 keV, no proton-decay branch could be identified [39]. Analysis of the predicted lifetimes for proton-decay paths to states in the daughter nuclide ^{157}Hf revealed that none of them would be sufficiently short to compete with the other decay branches. Indeed, in contrast to the case of ^{53}Co, the direct proton decay of the 19^- isomer to the $7/2^-$ ground state of ^{157}Hf would have a longer partial half-life than either of its lower-lying states.

4 Conclusions

Experimental studies of odd-Z nuclei have identified more than 30 proton-emitting nuclei, with examples known for most elements from iodine (Z = 53) to bismuth (Z = 83). These studies have revealed how the half-lives of low-lying states generally decrease with increasing neutron deficiency, dropping more rapidly when proton or alpha emission dominates over beta decay. For some elements it appears likely that the lightest isotopes whose ground states can be observed in conventional recoil separator experiments have already been reached.

The enhanced stability against proton emission of the high-lying isomers in ^{157,158}Ta [39] raises the possibility that multiparticle isomers could provide observable states in nuclei whose low-lying states decay by proton emission too quickly to be studied with a recoil separator. Calculations suggest that multiparticle isomers in neutron-rich nuclei [42, 43] and superheavy nuclei [44] may also have enhanced stability. Decays of long-lived isomers beyond the expected boundaries of the nuclear landscape are an intriguing prospect for future experiments.

Acknowledgements

This work was supported by the United Kingdom Science and Technology Facilities Council.

References

[1] P. M. Walker and G. D. Dracoulis, Nature (London) **399**, 35 (1999).
[2] K. P. Jackson *et al.*, Phys. Lett. B **33**, 281 (1970).
[3] J. Cerny *et al.*, Phys. Lett. B **33**, 284 (1970).
[4] S. Hofmann *et al.*, Z. Phys. A **305**, 111 (1982).
[5] O. Klepper *et al.*, Z. Phys. A **305**, 125 (1982).
[6] B. Blank and M. J. G. Borge, Prog. Part. Nucl. Phys. **60**, 403 (2008).
[7] L. W. Nordheim, Phys. Rev. **78**, 294 (1950).
[8] C. J. Gallgher Jr. and S. A. Moszkowski, Phys. Rev. **111**, 1282 (1958).
[9] R. J. Irvine *et al.*, Phys. Rev. C **55**, 1621 (1997).
[10] H. Kettunen *et al.*, Phys. Rev. C **69**, 054323 (2004).
[11] G. L. Poli *et al.*, Phys. Rev. C **59**, 2979 (1999).
[12] H. Kettunen *et al.*, Acta Phys. Pol. B **32**, 989 (2001).
[13] H. Mahmud *et al.*, Eur. Phys. J. A **15**, 85 (2002).
[14] C. N. Davids *et al.*, Phys. Rev. C **55**, 2255 (1997).
[15] M. C. Drummond *et al.*, Phys. Rev. C **89**, 064309 (2014).
[16] R. D. Page *et al.*, Phys. Rev. C **53**, 660 (1996).
[17] D. T. Joss *et al.*, Phys. Lett. B **641**, 34 (2006).
[18] R. D. Page *et al.*, Phys. Rev. C **75**, 061302 (2007).
[19] R. D. Page *et al.*, Phys. Rev. Lett. **68**, 1287 (1992).
[20] I. G. Darby *et al.*, Phys. Lett. B **695**, 78 (2011).
[21] I. G. Darby *et al.*, Phys. Rev. C **83**, 064320 (2011).
[22] P. J. Sapple *et al.*, Phys. Rev. C **84**, 054303 (2011).
[23] R. Grzywacz *et al.*, Eur. Phys. J. A **25**, 145 (2005).
[24] K. Livingston *et al.*, Phys. Lett. B **312**, 46 (1992).
[25] M. Karny *et al.*, Phys. Rev. Lett. **90**, 012502 (2003).
[26] A. P. Robinson *et al.*, Phys. Rev. C **68**, 054301 (2003).
[27] J. C. Batchelder *et al.*, Phys. Rev. C **57**, 1042 (1998).
[28] C. R. Bingham *et al.*, Phys. Rev. C **59**, 2984 (1999).
[29] T. N. Ginter *et al.*, Phys. Rev. C **61**, 014308 (1999).
[30] T. N. Ginter *et al.*, Phys. Rev. C **68**, 034330 (2003).
[31] P. J. Sellin *et al.*, Phys. Rev. C **47**, 1933 (1993).
[32] P. O. Larsson *et al.*, Z. Phys. A **314**, 9 (1983).
[33] R. D. Page, Phys. Rev. C **83**, 014305 (2011).
[34] H. Helppi *et al.*, Phys. Lett. B **115**, 11 (1982).
[35] J. H. McNeill *et al.*, Phys. Rev. Lett. **63**, 860 (1989).
[36] J. H. McNeill *et al.*, Z. Phys. A **344**, 3 (1993).
[37] S. Hofmann *et al.*, Z. Phys. A **291**, 53 (1979).
[38] S. Hofmann *et al.*, Z. Phys. A **333**, 107 (1989).
[39] R. J. Carroll *et al.*, Phys. Rev. Lett. **112**, 092501 (2014).
[40] R. J. Carroll *et al.*, Phys. Rev. C **93**, 034307 (2016).
[41] S. André *et al.*, Nucl. Phys. **A575**, 155 (1994).
[42] L. K. Peker *et al.*, Phys. Lett. B **36**, 547 (1971).
[43] V. P. Bugrov *et al.*, Sov. Jour. Nucl. Phys. **42**, 34 (1985).
[44] F. R. Xu *et al.*, Phys. Rev. Lett. **92**, 252501 (2004).

DOI: 10.1051/epjconf/201612302001

A dipole band above the I^π= 31/2$^-$ isomeric state in ^{189}Pb

D. M. Cullen[1,a]

[1] *School of Physics and Astronomy, University of Manchester, Manchester M13 9PL, United Kingdom*

Abstract. This contribution focuses on the new observation of a dipole band built upon an I^π=31/2$^-$ isomeric state in ^{189}Pb, identified using recoil-isomer tagging at the University of Jyväskylä, Finland. This is the lightest odd-mass Pb isotope in which a dipole band is known. By comparison with the heavier-mass dipole bands, the dipole band in ^{189}Pb was deduced to be based upon a $\pi(s_{1/2}^{-2}h_{9/2}i_{13/2})_{11^-} \otimes \nu(i_{13/2}^{-1})_{13/2^+}$ configuration. However, in the ^{189}Pb dipole band, the initial aligned angular momentum was larger than that exhibited by the dipole bands in the heavier-mass isotopes. This may be evidence for a reduced repulsive proton/neutron-hole interaction in ^{189}Pb.

1 Introduction

Dipole bands are a common feature of the neutron-deficient odd-mass Pb isotopes [1–8]. These bands exhibit a structure which resembles a strongly-coupled rotational band, however, the nuclear shape is approximately spherical and the strongly-coupled structure has strong $M1$- and weak $E2$-transitions.

Dipole bands in this region are known to originate from the coupling of protons in the $h_{9/2}$ orbital to neutron holes in the $i_{13/2}$ orbital [8–10]. The projections of the neutron and proton states are nearly perpendicular at the band-head because of the repulsive particle-hole interaction. Therefore, at the band-head the total angular momentum, I, points in a direction somewhere between the individual proton and neutron angular momentum. The higher-angular momentum states are formed by the step-wise alignment of the respective neutron and proton angular momentum with I in the shears mechanism [8, 10].

An isomeric state with a mean lifetime of 32 μs was first discovered in ^{189}Pb by Baxter *et al.* [11] but was not definitely linked into the lower-spin states. Subsequently, Dracoulis *et al.* were able to firmly link this state into the lower states and to define its spin and parity to be I^π=31/2$^-$ from electron conversion measurements at the Australian National University [9]. Dracoulis *et al.* also first proposed that this isomeric state was a shears-mode band-head state.

The present contribution describes the results of a new experiment which was performed at the University of Jyväskylä to establish the transitions which were built up on this isomeric state.

[a] e-mail: dave.cullen@manchester.ac.uk

2 Experiment

High-spin states were populated in ^{189}Pb with a ^{86}Kr beam at 355-MeV on a 500 μg/cm^2 ^{108}Pd target at the University of Jyväskylä. The beam current was ~10-pnA for a total of ~150 hours. The prompt gamma-rays were detected with the Jurogam-II detector array of 39 Compton suppressed germanium detectors with a total photo-peak efficiency of ~ 6% at 1.3 MeV. The recoiling nuclei passed through the RITU spectrometer [12] and implanted into a double-sided silicon strip detector. Subsequent gamma-ray decays were detected with the GREAT Planar- and Clover-focal plane germanium detectors [13]. All events were time stamped by the trigger-less data readout system [14]. The TDR data was used to perform recoil-isomer tagging by allowing the correlation of prompt Jurogam-II gamma-ray events at the target position with delayed GREAT Planar and Clover gamma-ray events at the focal plane. The full details of the experiment can be found in [15].

3 Results

In the analysis, several prompt-delayed matrices were constructed for events which were separated in time by up to 60 μs (~3 half-lives of the ^{189}Pb isomeric state). By gating on the known [9, 11] delayed gamma rays in ^{189}Pb, six new prompt transitions of energies 206-, 335-, 389-, 413-, 419- and 435-keV were established to lie above the 32 μs proposed dipole band-head state in ^{189}Pb. In addition, a prompt gamma-gamma matrix was constructed gated by delayed gamma rays detected up to 60μs (~3 half-lives of the ^{189}Pb isomeric state) after a recoil implant at the focal plane. This matrix was used to determine the coincidence relationships and placements of the gamma rays in the level scheme based on intensity measurements. In addition, the total intensity flow through particular levels was

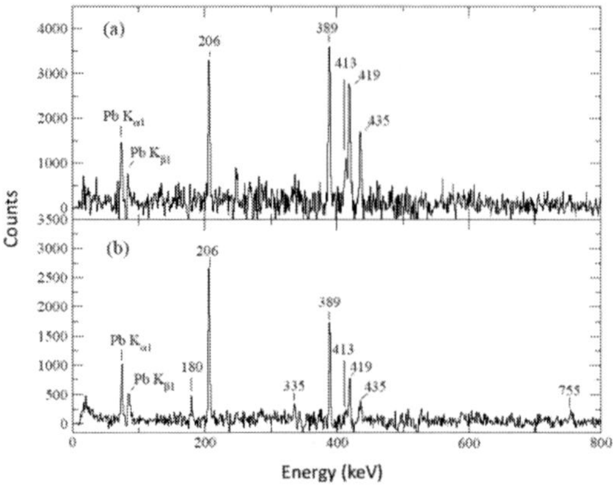

Figure 1. (a) Sum of prompt gates (206-, 389-, 419- and 435-keV) placed in a prompt gamma-gamma matrix gated by delayed gamma rays detected up to 60 μs (~3 half-lives of the ^{189}Pb isomeric state) after a recoil implant at the focal plane. (b) Prompt gamma-ray spectrum from a sum of delayed gates on the known transitions (142-, 267-, 272-, 530-, 609-, 811-, 819-, 911-, and 1142-keV) which depopulate the 32 μs isomeric state in ^{189}Pb [9, 11].

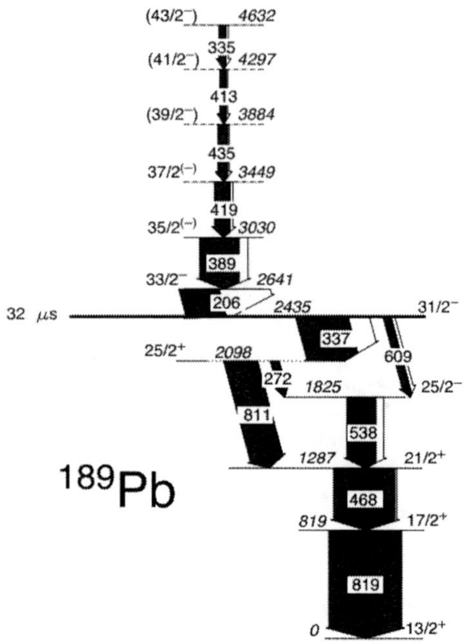

Figure 2. Partial level scheme of ^{189}Pb, showing the new transitions established in this work to form a dipole-band built upon the known 32 μs isomeric state [9].

used to estimate electron-conversion coefficients. The latter measurements along with angular correlation measurements were used to estimate the spins and parities of the levels. The full details of this analysis are published in Ref. [15]. Figure 1(a) shows a sum of individual prompt gates (206-, 389-, 419- and 435-keV) placed in this matrix. The labelled transitions are all in coincidence with each other and also the Pb Kα1 and Kβ1 X rays. Figure 1(b) shows a prompt gamma-ray spectrum gated by a sum of delayed gates on the known ^{189}Pb transitions (142, 267-, 272-, 530-, 609-, 811-, 819-, 911-, and 1142-keV) which depopulate the 32-μs isomeric state [9, 11] The spectrum shows the relative intensity of the newly established prompt transition sequence which feeds the isomeric state. Figure 1(b) shows the same sequence of gamma rays as Fig. 1(a) along with additional 180- and 755-keV transitions. These two transitions are not in coincidence with the main sequence and are thought to directly feed the isomeric state.

Figure 2 shows the sequence of transitions established in the present work to lie above the known 32 μs isomeric state [9, 11] in ^{189}Pb and the best estimates for the spins and parities of the levels

4 Discussion

In order to better interpret the structure of the newly established dipole band in ^{189}Pb, Figure 3 shows a comparison of the aligned angular momentum, i_x, as a function of rotational frequency for this band in comparison with the most intense dipole bands in the odd-mass ^{191}Pb-^{199}Pb isotopes [1, 3–5, 16]. Each band in the figure has been plotted assuming $K = 11$ and has a reference band with Harris Parameters $J_0 = 10\ \hbar^2$/MeV and $J_1 = 30\ \hbar^4$/MeV^3 subtracted in accordance with Ref. [1].

In this region, the dipole bands are known to be built upon a configuration at low rotational frequencies and a configuration at higher rotational frequencies above the first band crossing [1, 3–5, 16]. The first crossing takes place at rotational frequencies around 0.30-0.36 MeV and involves the alignment of two $i_{13/2}$ neutrons with a corresponding 8-9$\hbar$ gain of aligned angular momentum, see Fig. 3.

Figure 3 shows that the new dipole band in ^{189}Pb (filled circles) follows this general aligned angular momentum trend associated with the higher-mass neighbouring isotopes. However, the initial aligned angular momentum of the new dipole band in ^{189}Pb, like that of the dipole band in ^{191}Pb is ~2$\hbar$ larger than that of the neighbouring dipole bands. This increased initial aligned angular momentum in the two lightest mass dipole bands in the Pb nuclei, may suggest that the states in these nuclei are subject to a smaller repulsive proton/neutron-hole interaction. Such a reduced repulsive interaction could permit the angular momentum of the neutrons to contribute more to the total aligned angular momentum at low rotational frequencies. At higher rotational frequencies, this effect would become less noticeable as the proton and neutron-hole orbits align. A theoretical study of this effect would be necessary to fully interpret these differences.

DOI: 10.1051/epjconf/201612302001

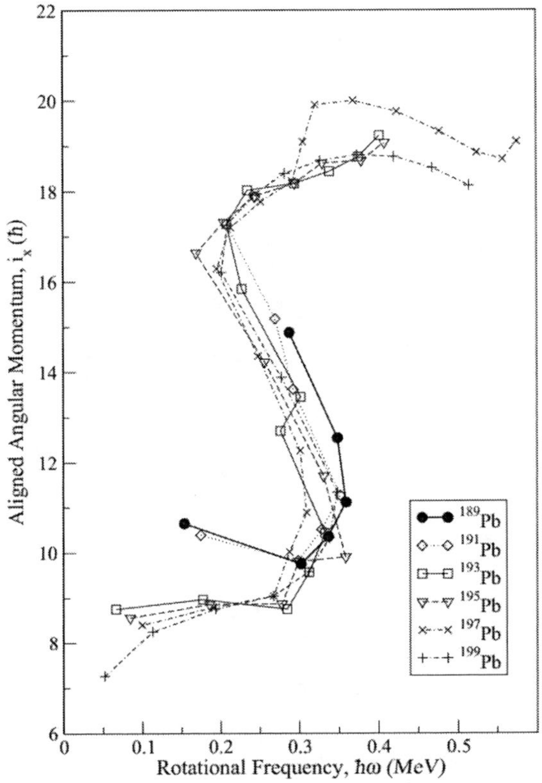

Figure 3. Aligned angular momentum systematic behaviour for the most intense dipole bands in the odd-mass $^{191-199}$Pb isotopes [1, 3–5, 16] along with the new dipole band in ^{189}Pb established in this work (filled circles). Note that the spins of the states in the dipole band in ^{191}Pb have been raised by $1\hbar$ in this figure.

Acknowledgements

I had the privilege of knowing, collaborating with and respecting George Dracoulis for nearly 20 years. This work on ^{189}Pb was initiated after discussions with George. Unfortunately George could not make it to the actual experiment but gave useful comments on the early stages of the work. Without George's original discovery of the $31/2^-$ isomer, this work would not have been possible. I remain indebted to and will miss him. This work was supported by the UK STFC through a standard Grant EP/E02551X/1.

References

[1] N. Fotiades *et al.*, Phys. Rev. C **57**, 1624 (1998).
[2] L. Ducroux *et al.*, Z. Phys. A **356**, 241 (1996).
[3] M. Kaci *et al.*, Z. Phys. A **354**, 267 (1996).
[4] A. Görgen *et al.*, Nucl. Phys. **A683**, 108 (2001).
[5] G. Baldsiefen *et al.*, Nucl. Phys. **A574**, 521 (1994).
[6] G. D Dracoulis, A. P. Byrne, and A. Baxter, Phys. Lett. B **432**, 37 (1998).
[7] G. D. Dracoulis *et al.*, Phys. Rev. C **67**, 051301(R) (2003).
[8] R. M. Clark and A. O. Macchiavelli, Annu. Rev. Nucl. Part. Sci. **50**, 1 (2000).
[9] G. D. Dracoulis, G. J. Lane, T. Kibédi, and P. Nieminen, Phys. Rev. C **79**, 031302 (2009).
[10] H. Hübel, Prog. Part. Nucl. Phys. **54**, 1 (2005).
[11] A. M. Baxter *et al.*, Phys. Rev. C **71**, 054302 (2005).
[12] M. Leino *et al.*, Nucl. Instrum. Methods B **99**, 653 (1995).
[13] R. D. Page *et al.*, Nucl. Inst. Meth. B **204**, 634 (2003).
[14] I. Lazarus *et al.*, IEEE Trans. Nucl. Sci. **48**, 567 (2001).
[15] D. Hodge *et al.*, Phys. Rev. C **92**, 054312 (2015).
[16] D. L. Balabanski *et al.*, Phys. Rev. C **83**, 014304 (2011).

EPJ Web of Conferences **123**, 02002 (2016)
Heavy Ion Accelerator Symposium 2015

DOI: 10.1051/epjconf/201612302002

β-decay spectroscopy of neutron-rich ^{160,161,162}Sm isotopes

Z. Patel[1,2,a], Zs. Podolyák[1], P. M. Walker[1], P. H. Regan[1,3], P.-A. Söderström[2], H. Watanabe[2,4,5], E. Ideguchi[6,7], G. S. Simpson[8], S. Nishimura[2], F. Browne[2,10], P. Doornenbal[2], G. Lorusso[2,3], S. Rice[1,2], L. Sinclair[2,11], T. Sumikama[12], J. Wu[2,9], Z. Y. Xu[13], N. Aoi[6,7], H. Baba[2], F. L. Bello Garrote[14], G. Benzoni[15], R. Daido[7], Zs. Dombrádi[22], Y. Fang[7], N. Fukuda[2], G. Gey[8], S. Go[16], A. Gottardo[17], N. Inabe[2], T. Isobe[2], D. Kameda[2], K. Kobayashi[18], M. Kobayashi[16], T. Komatsubara[19,20], I. Kojouharov[21], T. Kubo[2], N. Kurz[21], I. Kuti[22], Z. Li[23], H. L. Liu[24], M. Matsushita[16], S. Michimasa[16], C.-B. Moon[25], H. Nishibata[7], I. Nishizuka[12], A. Odahara[7], E. Şahin[14], H. Sakurai[2,13], H. Schaffner[21], H. Suzuki[2], H. Takeda[2], M. Tanaka[7], J. Taprogge[26,27], Zs. Vajta[22], F. R. Xu[9], A. Yagi[7], and R. Yokoyama[16]

[1] *Department of Physics, University of Surrey, Guildford, GU2 7XH, United Kingdom*
[2] *RIKEN Nishina Center, 2-1 Hirosawa, Wako-shi, Saitama 351-0198, Japan*
[3] *National Physical Laboratory, Teddington, Middlesex, TW11 0LW, United Kingdom*
[4] *International Research Center for Nuclei and Particles in the Cosmos, Beihang University, Beijing 100191, China*
[5] *School of Physics and Nuclear Energy Engineering, Beihang University, Beijing 100191, China*
[6] *Research Center for Nuclear Physics (RCNP), Osaka University, Ibaraki, Osaka 567-0047, Japan*
[7] *Department of Physics, Osaka University, Machikaneyama-machi 1-1, Osaka 560-0043 Toyonaka, Japan*
[8] *LPSC, Université Joseph Fourier Grenoble 1, CNRS/IN2P3, Institut National Polytechnique de Grenoble, F-38026 Grenoble Cedex, France*
[9] *School of Physics and State Key Laboratory of Nuclear Physics and Technology, Peking University, Beijing 100871, China*
[10] *School of Computing, Engineering and Mathematics, University of Brighton, Brighton, BN2 4JG, United Kingdom*
[11] *Department of Physics, University of York, Heslington, York, YO10 5DD, United Kingdom*
[12] *Department of Physics, Tohoku University, Aoba, Sendai, Miyagi 980-8578, Japan*
[13] *Department of Physics, University of Tokyo, Hongo, Bunkyo-ku, Tokyo 113-0033, Japan*
[14] *Department of Physics, University of Oslo, Oslo, Norway*
[15] *INFN Sezione di Milano, I-20133 Milano, Italy*
[16] *Center for Nuclear Study (CNS), University of Tokyo, Wako, Saitama 351-0198, Japan*
[17] *Instituto Nazionale di Fisica Nucleare, Laboratori Nazionali di Legnaro, I-35020 Legnaro, Italy*
[18] *Department of Physics, Rikkyo University, 3-34-1 Nishi-Ikebukuro, Toshima-ku, Tokyo 171-8501, Japan*
[19] *Research Facility Center for Pure and Applied Science, University of Tsukuba, Ibaraki 305-8577, Japan*
[20] *Rare Isotope Science Project, Institute for Basic Science, Daejeon 305-811, Korea*
[21] *GSI Helmholtzzentrum für Schwerionenforschung GmbH, 64291 Darmstadt, Germany*
[22] *Institute for Nuclear Research, Hungarian Academy of Sciences, P. O. Box 51, Debrecen, H-4001, Hungary*
[23] *School of Physics, Peking University, Beijing 100871, China*
[24] *Hoseo University, Asan, Chungnam 336-795, Korea*
[25] *Instituto de Estructura de la Materia, CSIC, E-28006 Madrid, Spain*
[26] *Departamento de Física Teórica, Universidad Autónoma de Madrid, E-28049 Madrid, Spain*

Abstract. Neutron-rich ^{160,161,162}Sm isotopes have been populated at the RIBF, RIKEN via β decay for the first time. β-coincident γ rays were observed in all three isotopes including γ rays from the isomeric decay of ^{160}Sm and ^{162}Sm. The isomers in ^{160}Sm and ^{162}Sm have previously been observed but have been populated via β decay for the first time. The isomeric state in ^{162}Sm is assigned a $4^-\ \nu\frac{7}{2}^+[633]\otimes\nu\frac{1}{2}^-[521]$ configuration based on the decay pattern. The level schemes of ^{160}Sm and ^{162}Sm are presented. The ground states in the parent nuclei ^{160}Pm and ^{162}Pm are both assigned a $6^-\ \nu\frac{7}{2}^+[633]\otimes\pi\frac{5}{2}^-[532]$ configuration based on the population of states in the daughter nuclei. Blocked BCS calculations were performed to further investigate the spin-parities of the ground states in ^{160}Pm, ^{161}Pm, and ^{162}Pm, and the isomeric state in ^{162}Sm.

1 Introduction

The β decay of neutron-rich nuclei can be a useful tool to probe low-lying states in nuclei that are otherwise difficult to populate. Here we present nuclear data on ^{160}Sm, ^{161}Sm, and ^{162}Sm that have been populated via β decay for the first time. These nuclei lie in the neutron-rich mid-shell region between the closed shells of $Z = 50$ and 82 and $N = 82$ and 126 and are expected to be among the most col-

[a] e-mail: z.patel@surrey.ac.uk

DOI: 10.1051/epjconf/201612302002

lective of nuclei. Therefore this is a good testing ground for collective models. Neutron-rich nuclei are created in the *r* process, which is responsible for the production of half the nuclei that exist with $Z > 26$ (iron). There are three major peaks in the elemental abundances at $A \approx 80$, 130, and 195. These are due to the closed shells at $N = 50$, 82, and 126 that create "waiting" points in the *r* process [1]. Experimental data has led to an improved understanding of these peaks [2–4]. There is also a less pronounced peak at $A \approx 160$, of which the origin is not as clear. It is thought to be due to a deformation maximum [5], however recent theories [6, 7] and experimental data [8] have also suggested that a subshell gap at $N = 100$ may play a role. Investigation of nuclei around $A = 160$ may provide insight into the possible origins of this peak. The high intensity uranium beam at the RIBF, RIKEN has been utilised to populate these isotopes at the limits of known nuclei.

2 Experimental method

Neutron-rich ^{160,161,162}Pm isotopes were produced by in-flight fission of a 345 A·MeV ^{238}U beam with an average beam intensity of 10 particle-nA incident on a ^{9}Be target at the RIBF. The secondary beam was passed through BigRIPS and the ZeroDegree spectrometers [9, 10] where the nuclei were separated according to their mass-to-charge ratio (A/q) and atomic number (Z) using time-of-flight, magnetic rigidity and energy loss (TOF, Bρ, and ΔE) [11, 12].

The ions were implanted into an active stopper, WAS3ABi: Wide Angle Silicon-Strip-Stopper Array for Beta and ion detection [13]. This consisted of 5 DSSDs, each with 60 x 40 1-mm strips. A time window of 1 second was used to correlate an ion implantation with a β-decay particle detected in the same or neighbouring pixel. The γ rays emitted following β decay were detected using EURICA (Euroball-RIKEN Cluster Array) [14–16]: 84 HPGe crystals arranged in a 4π configuration at $\sim$ 22 cm from ion and β implantation.

3 Results

A γ-ray energy-time matrix was produced for each Sm nucleus by placing a particle gate on the parent Pm nucleus. Four Pm nuclei were populated: ^{160}Pm, ^{161}Pm, ^{162}Pm, and ^{163}Pm. To study the γ rays in prompt coincidence with the β decay of the parent nucleus a short time gate of < 500 ns was used. A time gate of ≥ 500 ns was used to investigate possible isomeric population following β decay. Four Pm isotopes were populated in this experiment, however the number of implanted ^{163}Pm ions is much lower than the A = 160-162 nuclei and no γ rays were observed following the decay of this nucleus. Prompt coincident γ rays associated with the β decay of ^{160}Pm, ^{161}Pm, and ^{162}Pm can be seen in Fig. 1, Fig. 3, and Fig. 4 respectively.

3.1 ^{160}Sm

Four γ rays of 108, 162, 250, and 1128 keV known to belong to the decay of ^{160}Sm are seen in Fig. 1 [17, 18].

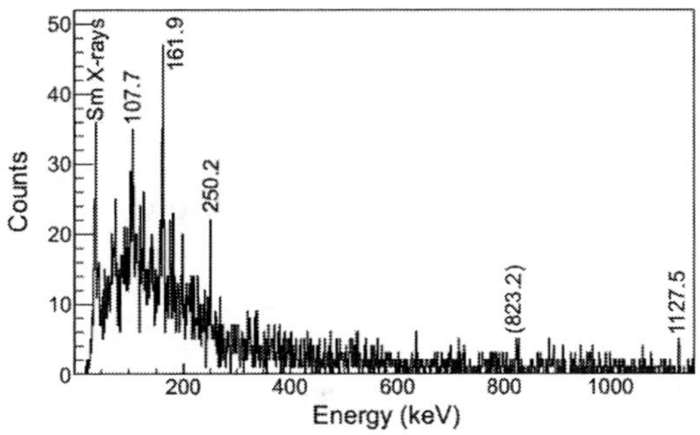

Figure 1. Energy spectrum of prompt γ rays in ^{160}Sm.

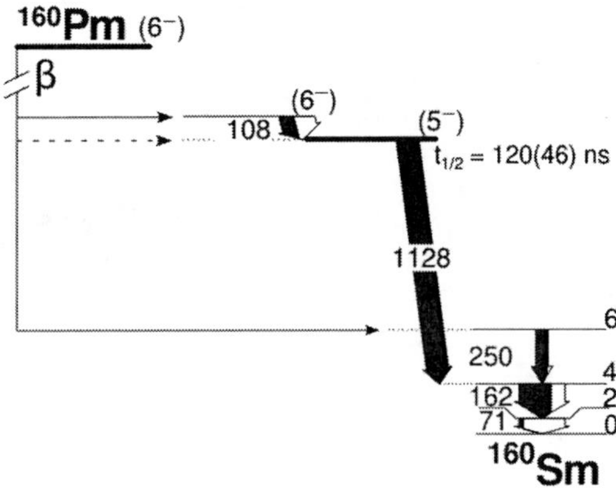

Figure 2. Level scheme of ^{160}Sm populated by the β decay of ^{160}Pm.

There is also a γ-ray peak visible at 823 keV, however, the origin of this γ ray is unknown. The 162 and 250 keV γ rays belong to the $4^+ \rightarrow 2^+$ and $6^+ \rightarrow 4^+$ transitions in the ground state band respectively. The 1128 keV γ ray is from the decay of a (5$^-$) $\pi\frac{5}{2}^-[532] \otimes \pi\frac{5}{2}^+[413]$ isomeric state [18] with a half-life of 120(46) ns [17]. The 108 keV γ ray originates from the (6$^-$) to (5$^-$) state transition [18].

The population of states in ^{160}Sm via the β decay of ^{160}Pm are shown in the level scheme in Fig. 2. The tentative assignment of a (6$^-$) spin-parity to the ground state of the parent nucleus is made on the basis that the β decay populates the (6$^-$) and 6$^+$ states in ^{160}Sm. The intensities of the levels, measured in a 0 – 600 ns time window to include 5 half-lives of the isomeric state are shown in Tab. 1. The intensities balance through the levels within statistical uncertainties. It is likely that the β decay of ^{160}Pm populates the (5$^-$) isomeric state directly, however there is no direct experimental evidence for this such as missing intensity, hence this transition is indicated by a dashed line in Fig. 2. Blocked BCS calculations presented in Table 3 also support a (6$^-$) spin-parity assignment of the ^{160}Pm ground state.

DOI: 10.1051/epjconf/201612302002

Table 1. Initial level energy, E_i and spin-parity, J_i^π of the levels of ^{160}Sm in this work. For each γ ray the energy, E_γ, γ-ray intensity, I_γ relative to the 162-keV γ-ray intensity, and final level spin, J_f^π, are listed.

E_i (keV)	J_i^π	E_γ (keV)	$I_\gamma(rel.)$	J_f^π
71(1)	(2$^+$)	71(1)[a]		0$^+$
233(1)	(4$^+$)	161.9(5)	100	(2$^+$)
483(1)	(6$^+$)	250.2(5)	35(18)	(4$^+$)
1361(1)	(5$^-$)	1127.5(5)	37(10)	(4$^+$)
1469(1)	(6$^-$)	107.7(8)	22(10)	(5$^-$)

[a] The 71-keV γ ray is only observed in coincident spectra.

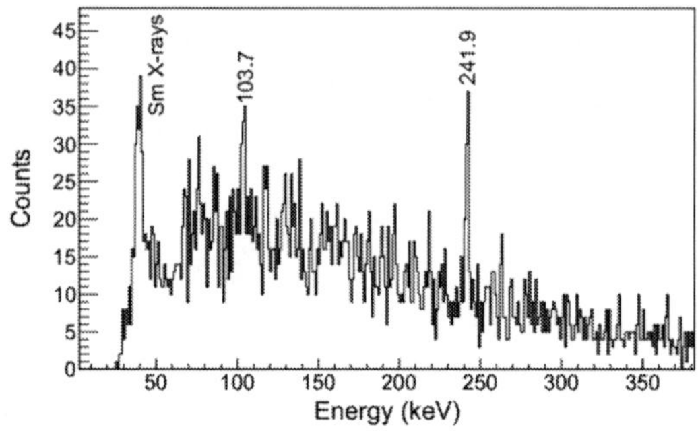

Figure 3. Energy spectrum of prompt γ rays in ^{161}Sm.

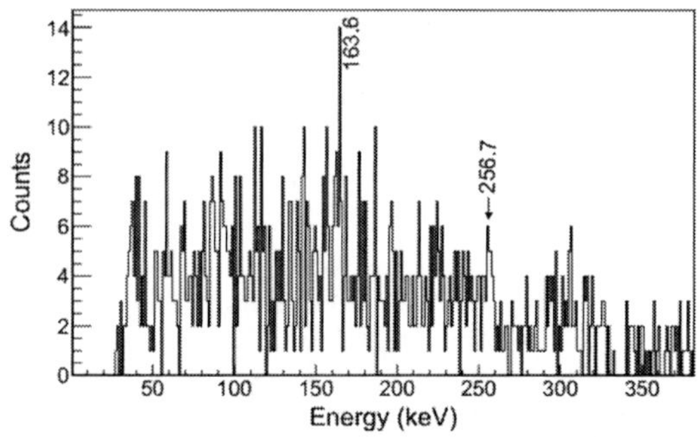

Figure 4. Energy spectrum of prompt γ rays in ^{162}Sm.

3.2 ^{161}Sm

Two γ rays of 104-and 242-keV energy are seen in prompt coincidence with the β decay of ^{161}Pm, shown in Fig. 3. They do not appear in coincidence with each other, or in a delayed $\beta-\gamma$ coincident spectrum, therefore it is likely they belong to two different bands, each populated directly by β decay. Blocked BCS calculations presented in Section 4.2 show several possiblilies for the spin-parities of the states in ^{161}Pm and ^{161}Sm.

3.3 ^{162}Sm

Two γ rays are visible in prompt coincidence with the β decay of ^{162}Pm into ^{162}Sm. From systematics of neighbouring nuclei it is likely that the 164-keV γ ray is from the $4^+ \to 2^+$ transition and the 257-keV γ ray is from the $6^+ \to 4^+$ transition. The $2^+ \to 0^+$ transition is expected to be around 70 keV from systematics. It was not observed due to the high internal conversion rate expected for this low energy transition.

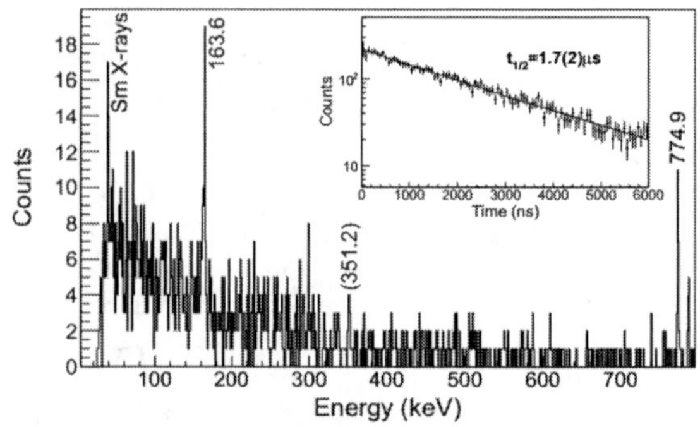

Figure 5. β-delayed γ energy spectrum of ^{162}Sm with t ≥ 500 ns. A 789-keV γ ray from the self activity in the $LaBr_3$(Ce) detectors added to EURICA is also visible. Inset: The exponential decay curve is from the isomeric decay of ^{162}Sm, found using the passive stopper part of the experiment.

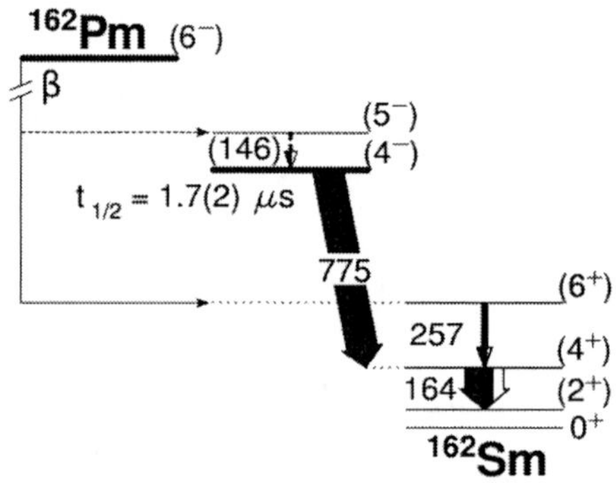

Figure 6. Level scheme of ^{162}Sm populated by the β decay of ^{162}Pm.

Table 2. Initial level energy, E_i and spin-parity, J_i^π of the levels of ^{162}Sm in this work. For each γ ray the energy, E_γ, γ-ray intensity, I_γ relative to the 775-keV γ-ray intensity, and final level spin, J_f^π, are listed.

$E_i - E(2^+)$ (keV)	J_i^π	E_γ (keV)	$I_\gamma(rel.)$	J_f^π
163.6(7)	(4$^+$)	163.6(7)	87(29)	(2$^+$)
420(1)	(6$^+$)	256.7(9)	15(3)	(4$^+$)
938.5(8)	(4$^-$)	774.9(5)	100	(4$^+$)
1085(1)	(5$^-$)	146(1)[a]		(4$^-$)

[a] The 146-keV γ ray is only observed in coincident spectra.

An isomeric state in ^{162}Sm is populated via β decay, shown in the β-delayed coincident γ-ray spectrum in Fig. 5. A 775-keV γ ray appears and the 257-keV γ ray is no longer visible in the delayed γ-ray spectrum which suggests the β decay populates an isomeric state of ^{162}Sm with

DOI: 10.1051/epjconf/201612302002

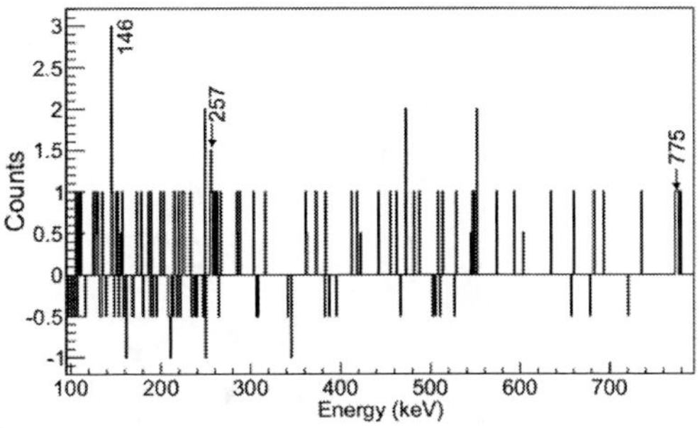

Figure 7. γ-ray spectrum gated on the 164-keV γ ray. A 146-keV γ ray is visible in addition to the 257 and 775 keV γ rays.

a half-life longer than 500 ns in addition to the (6^+) state in the ground band. The 775-keV transition directly feeds the 164-keV transition, therefore it populates the (4^+) ground band state. The spin-parity of the isomeric state is assigned as (4^-) based on a lack of transitions to the (2^+) and (6^+) ground band states. This is supported by blocked BCS calculations presented in Table 4. The level scheme of ^{162}Sm is shown in Fig. 6. The intensities of the γ rays, shown in Table 2, are calculated at $t = 0$ using the exponential law of radioactive decay, $N(t) = N_0 e^{(-\lambda t)}$, in order to compare the intensities of the isomer-delayed γ rays to the intensities of the prompt γ rays.

The spin-parity of the ground state in the parent nucleus, ^{162}Pm, is assigned (6^-) due to the decay to the (6^+) ground band state, similar to ^{160}Sm. Blocked BCS calculations in Table 3 support this assignment. However, the (6^-) state in ^{162}Pm is strongly forbidden to decay into the (4^-) isomeric state in ^{162}Sm due to the $\Delta I = 2$ and parity change required. Therefore it is likely that the ^{162}Pm ground state decays into a level on top of the isomeric state. Indeed, a 146-keV γ ray is visible in a background subtracted 164-keV coincidence spectrum, shown in Fig. 7. This γ ray is tentatively placed as a transition from a (5^-) state to the isomeric (4^-) state in Fig. 6.

These Sm nuclei were also produced directly in the fission reaction at the start of BigRIPS. In order to study isomers in these nuclei a passive stopper was used in place of WAS3ABi, which allowed for a higher implantation rate. The experimental set-up is detailed in [8, 19]. The (4^-) isomeric state in ^{162}Sm was also observed in this passive stopper part of the experiment, with a measured high-precision half-life of 1.7(2) μs [19], seen in Fig. 5.

4 Discussion

Blocked BCS calculations of the type in Ref. [20] were performed in order to better understand the spin-parities of the ground states in ^{160}Pm, ^{161}Pm, and ^{162}Pm, and to understand the spin-parity of the isomeric state in ^{162}Sm. The pairing strengths were fixed as $G_\pi = 21.0\,A\cdot$MeV and $G_\nu = 20.0\,A\cdot$MeV, in accordance with Jain *et al.* [20], and the deformation parameters were taken from Möller *et al* [21]. The results can be seen in Table 3 and Table 4.

Table 3. Low-lying quasiparticle states in ^{160}Pm, ^{161}Pm, and ^{162}Pm predicted by blocked-BCS calculations. Only K^π values of favoured spin-couplings are shown.

Nucleus	K^π	Configuration	E_x (keV)
^{160}Pm	6^-	$\nu\frac{7}{2}^+[633] \otimes \pi\frac{5}{2}^-[532]$	0
^{160}Pm	2^+	$\nu\frac{1}{2}^-[521] \otimes \pi\frac{5}{2}^-[532]$	52
^{160}Pm	3^-	$\nu\frac{1}{2}^-[521] \otimes \pi\frac{5}{2}^+[413]$	289
^{160}Pm	5^+	$\nu\frac{5}{2}^-[512] \otimes \pi\frac{5}{2}^-[532]$	436
^{161}Pm	$\frac{5}{2}^-$	$\pi\frac{5}{2}^-[532]$	0
^{161}Pm	$\frac{5}{2}^+$	$\pi\frac{5}{2}^-[413]$	268
^{161}Pm	$\frac{3}{2}^+$	$\pi\frac{3}{2}^+[411]$	586
^{161}Pm	$\frac{3}{2}^-$	$\pi\frac{3}{2}^-[541]$	672
^{162}Pm	2^+	$\nu\frac{1}{2}^-[521] \otimes \pi\frac{5}{2}^-[532]$	0
^{162}Pm	6^-	$\nu\frac{7}{2}^+[633] \otimes \pi\frac{5}{2}^-[532]$	44
^{162}Pm	5^+	$\nu\frac{5}{2}^-[512] \otimes \pi\frac{5}{2}^-[532]$	163
^{162}Pm	3^-	$\nu\frac{1}{2}^-[521] \otimes \pi\frac{5}{2}^-[413]$	196

4.1 ^{160}Sm

The lowest energy configuration for the ground state of ^{160}Pm from blocked BCS calculations presented in Table 3 is 6^- $\nu\frac{7}{2}^+[633] \otimes \pi\frac{5}{2}^-[532]$. This supports the assignment made in Section 3.1. The ground state of ^{160}Pm is assigned a (6^-) spin-parity that β decays to the (6^-) and (6^+) states in ^{160}Sm. A 2^+ $\nu\frac{1}{2}^-[521] \otimes \pi\frac{5}{2}^-[532]$ configuration is also predicted by the calculations to have low energy. It is possible that a second state with 2^+ spin-parity exists in ^{160}Pm that decays to the 2^+ ground band state in ^{160}Sm, however, due to lack of experimental evidence the 2^+ state in ^{160}Pm has not been placed in the level scheme. The reduced hindrance of the 1128-keV transition depopulating the 5^- isomer is $f_\nu \sim 20$. The reduced hindrances of the two known K isomers in ^{160}Sm are discussed in detail in reference [18].

4.2 ^{161}Sm

Blocked BCS calculations display many possible ground state configurations of ^{161}Pm shown in Table 3. The calculations also reveal many possible low-lying states in ^{161}Sm. The two γ rays in the energy spectrum of ^{161}Sm are not in coincidence with each other. It is possible that they arise from the β decay of two different states in ^{161}Pm into a single state in ^{161}Sm. It is also possible that they are both populated from the β decay of one state in ^{161}Pm into two different states in ^{161}Sm. Therefore it is not possible to determine the decay pattern or to assign spin-parities to these possible states.

4.3 ^{162}Sm

The blocked BCS calculations for the ground state of ^{162}Pm show a similar result to the calculations for ^{160}Pm. A 2^+ $\nu\frac{1}{2}^-[521] \otimes \pi\frac{5}{2}^-[532]$ configuration is lowest in energy, followed by a 6^- $\nu\frac{7}{2}^+[633] \otimes \pi\frac{5}{2}^-[532]$ configuration

Table 4. Low-lying quasiparticle states in ^{162}Sm predicted by blocked-BCS calculations.

K^{π}	configuration	E_x (MeV)
5^-	$\pi\frac{5}{2}^-[532] \otimes \pi\frac{5}{2}^-[413]$	1.000
4^-	$\nu\frac{7}{2}^+[633] \otimes \nu\frac{1}{2}^-[521]$	1.043
4^-	$\pi\frac{5}{2}^-[532] \otimes \pi\frac{3}{2}^+[411]$	1.547 ([a])
4^+	$\pi\frac{5}{2}^-[413] \otimes \pi\frac{3}{2}^+[411]$	1.614
3^+	$\nu\frac{1}{2}^-[521] \otimes \nu\frac{5}{2}^-[512]$	1.668
6^-	$\nu\frac{7}{2}^+[633] \otimes \nu\frac{5}{2}^-[512]$	1.797 ([a])

[a] 200 keV has been added to these states as they have energetically unfavoured configurations according to the residual spin-spin coupling rule.

only 44 keV apart. The ground state in ^{162}Pm is assigned a (6^-) spin-parity due to population of the (6^+) ground band state in ^{162}Sm. The 257-keV transition from the (6^+) state is weakly populated compared to the isomeric state, therefore it is determined to be a forbidden transition. The decay to the (5^-) state on top of the isomeric state is allowed. It is possible that a second ground state in ^{162}Pm exists with a spin-partiy of 2^+ as in ^{160}Pm, however experimental evidence is not present. If it exists it would decay into the (2^+) ground band state in ^{162}Sm.

The lowest energy quasiparticle configurations in ^{162}Sm predicted by blocked BCS calculations are a 5^- $\pi\frac{5}{2}^-[532] \otimes \pi\frac{5}{2}^-[413]$ configuration and a 4^- $\nu\frac{7}{2}^+[633] \otimes \nu\frac{1}{2}^-[521]$ that are only 43 keV apart. The isomeric state in ^{162}Sm is assigned a (4^-) spin-parity based on the lack of observation of a transition from the isomeric state to the (6^+) state. If such a transition existed it would be visible in the γ-ray spectrum. This assignment is also supported by the calculations presented in Table 4.

The reduced hindrance of the 775-keV transition depopulating the 4^- isomer is $f_\nu \sim 70$. This is larger than in ^{160}Sm, but consistent with the limit found for the 4^- isomer in ^{154}Nd [17].

5 Conclusions

Low-lying states in ^{160}Sm, ^{161}Sm, and ^{162}Sm were populated via β decay for the first time. Isomeric states in ^{160}Sm and ^{162}Sm were also observed following β decay, assigned a (5^-) $\pi\frac{5}{2}^-[532] \otimes \pi\frac{5}{2}^+[413]$ [18] and a (4^-) $\nu\frac{7}{2}^+[633] \otimes \nu\frac{1}{2}^-[521]$ configuration respectively. The ground states in ^{160}Pm and ^{162}Pm were determined to have a (6^-) $\nu\frac{7}{2}^+[633] \otimes \pi\frac{5}{2}^-[532]$ configuration. Further experimental evidence is required to determine the ground state of ^{161}Pm. Further work is also required to extract the β decay half-life and the Q_β values of the parent nuclei ^{160}Pm, ^{161}Pm, and ^{162}Pm.

Acknowledgements

This work was carried out at the RIBF operated by RIKEN Nishina Center, RIKEN and CNS, University of Tokyo. All UK authors are supported by STFC. P. H. R. and G. L. are partially supported by the UK National Measurement Office (NMO). P.-A. S. was financed by JSPS Grant Number 23.01752 and the RIKEN Foreign Postdoctoral Researcher Program. J. T. was financed by Spanish Ministerio de Ciencia e Innovación under Contracts No. FPA2009-13377-C02 and No. FPA2011-29854-C04. Zs. D., I. K., and Zs. V. were supported by OTKA contract number K100835. We acknowledge the EUROBALL Owners Committee for the loan of germanium detectors and the PreSpec Collaboration for the readout electronics of the cluster detectors. This work was supported by JSPS KAKENHI Grant No. 25247045.

References

[1] E. M. Burbidge, G. R. Burbidge, W. A. Fowler *et al.*, Rev. Mod. Phys. **29**, 547 (1957).
[2] J. A. Winger, S. V. Ilyushkin, K. P. Rykaczewski *et al.*, Phys. Rev. Lett. **102**, 142502 (2009).
[3] A. Jungclaus, L. Cáceres, M. Górska *et al.*, Phys. Rev. Lett. **99**, 132501 (2007).
[4] A. I. Morales, J. Benlliure, T. Kurtukián-Nieto *et al.*, Phys. Rev. Lett. **113**, 022702 (2014).
[5] R. Surman, J. Engel, J. R. Bennett *et al.*, Phys. Rev. Lett. **79**, 1809 (1997).
[6] S. K. Ghorui, B. B. Sahu, C. R. Praharaj *et al.*, Phys. Rev. C **85**, 064327 (2012).
[7] S. K. Ghorui (2014)., private communication
[8] Z. Patel, P. A. Söderström, Z. Podolyák *et al.*, Phys. Rev. Lett. **113**, 262502 (2014).
[9] Y. Yano, Nucl. Instrum. Meth. B **261**, 1009 (2007).
[10] T. Kubo, D. Kameda, H. Suzuki *et al.*, Prog. Theor. Exp. Phys. p. 03C003 (2012).
[11] T. Kubo, D. Kameda, H. Suzuki *et al.*, Prog. Theor. Exp. Phys. p. 03C003 (2012).
[12] Y. Yano, Nucl. Instrum. Meth. B **261**, 1009 (2007).
[13] S. Nishimura, G. Lorusso, Z. Xu *et al.*, RIKEN Accel. Prog. Rep. **46**, 182 (2013).
[14] S. Nishimura, Nucl. Phys. News Int. **22**, 38 (2012).
[15] P. A. Söderström, S. Nishimura, P. Doornenbal *et al.*, Nucl. Instrum. Meth. B **317**, 649 (2013).
[16] P. A. Söderström, S. Nishimura, P. Doornenbal *et al.*, Nucl. Instrum. Meth. B **317**, 649 (2013).
[17] G. S. Simpson, W. Urban, J. Genevey *et al.*, Phys. Rev. C **80**, 024304 (2009).
[18] Z. Patel, Z. Podolyák, P. Walker *et al.*, Phys. Lett. B **753**, 182 (2016).
[19] Z. Patel, Ph.D. thesis, The University of Surrey (2015).
[20] K. Jain, O. Burglin, G. Dracoulis *et al.*, Nucl. Phys. **A591**, 61 (1995).
[21] P. Möller, J. Nix, W. Myers *et al.*, Atom. Data Nucl. Data **59**, 185 (1995).

EPJ Web of Conferences **123**, 02003 (2016) DOI: 10.1051/epjconf/201612302003
Heavy Ion Accelerator Symposium 2015

Towards Superheavies: Spectroscopy of 94 < Z < 98, 150 < N < 154 Nuclei

P. Chowdhury[1,a], S. S. Hota[1], Y, Qiu[1], I. Ahmad[2], M. P. Carpenter[2], J. P. Greene[2], R. V. F. Janssens[2], T. L. Khoo[2], F. G. Kondev[2], T. Lauritsen[2], C. J. Lister[1], D. Seweryniak[2], and S. Zhu[2]

[1] *Department of Physics and Applied Physics, University of Massachusetts Lowell, Lowell, Massachusetts 01854, USA*
[2] *Argonne National Laboratory, Argonne, Illinois 60439, USA*

Abstract. The heaviest nuclei where excitations above the ground state can be studied lie near $Z \sim 100$. These nuclear structure studies are important testing grounds for theoretical models that aim to describe superheavy nuclei. To study the highest neutron orbitals ($150 \leq N \leq 154$), we have populated high angular momentum states in a series of Pu ($Z = 94$), Cm ($Z = 96$) and Cf ($Z = 98$) nuclei, via inelastic and transfer reactions, with heavy beams on long-lived radioactive actinide targets. Multiple collective excitation modes and structures were identified, and their configurations deduced. Quasiparticle alignments are mapped, with odd-A band structures helping identify specific orbital contributions via blocking arguments. Higher-order multipole shapes are observed to play a significant role in disentangling competing neutron and proton alignments. The $N > 152$ data provide new perspectives on physics beyond the $N = 152$ sub-shell gap.

1 Introduction

The structure of nuclei that lie at the edges of stability in the nuclear landscape hold the most discovery potential for new physics. While nuclei along the proton and neutron drip-lines reveal stellar nucleosynthesis pathways, very heavy high-Z nuclei, whose fragility stems from enhanced Coulomb repulsion, lead us towards superheavy physics. The island of superheavy nuclei predicted to lie at the next doubly magic proton and neutron shell-gap is a topic of intense interest in contemporary nuclear structure research. While steady, albeit slow, progress is being made towards synthesising superheavy elements, their picobarn production cross-sections via fusion preclude the possibility of studying excitations built on them for some time to come. The heaviest nuclei where such spectroscopy is possible lie near $Z \sim 100$ [1], where the nuclei exhibit surprisingly robust fission barriers up to high angular momenta [2]. These studies provide critical input for constraining theoretical models that attempt to describe the physics of superheavy nuclei, which include single-particle energies, shell gaps and pairing. While $Z \geq 100$ nuclei can be produced and studied via fusion-evaporation reactions, we have concentrated on $Z < 100$ nuclei, where inelastic and transfer reactions are possible, as these are the heaviest long-lived radioactive nuclei which can be used as targets. Compared to fusion reactions leading to $Z \geq 100$, inelastic and transfer reactions with $Z < 100$ nuclei have comparatively higher cross-sections, and can populate more neutron-rich nuclei. We have focused on studying the highest neutron orbitals with $150 \leq N \leq 154$. These studies follow on the successes of prior investigations in this region using similar techniques [3–5].

[a] e-mail: partha_chowdhury@uml.edu

2 Experiments

In a series of experiments, we have populated high angular momentum states in a range of nuclei: $^{244-246}$Pu ($Z = 94$), $^{245-250}$Cm ($Z = 96$) and $^{248-251}$Cf ($Z = 98$). Beams of ^{207,208}Pb and ^{209}Bi from the ATLAS accelerator facility at Argonne were used to bombard backed radioactive targets of ^{244}Pu, ^{248}Cm and $^{249-251}$Cf, with beam energies ~15% above the Coulomb barrier. The gamma rays were detected by ~100 Compton-suppressed Ge detectors of the Gammasphere array. The radioactive targets were deposited on a ~50-mg/cm^2 ^{197}Au backing and sealed with a thin (~150 μg/cm^2) layer of Au in front. The target activity and strong Coulomb excitation of the Au backing pose considerable experimental challenges in optimising and monitoring beam-on-target conditions. Extracting the relevant spectra from this overwhelming background required the full power of the array, both in solid angle and granularity, to extract spectroscopic information.

To illustrate this point, a summed spectrum of double-coincidence gates on transitions in the ground state band of the target nucleus ^{244}Pu is first compared in Fig. 1 to the ungated total projection of the γ-γ-γ cube. In this case, the ^{244}Pu target activity was low ($\approx$1 nCi), and the total projection spectrum is dominated by the Coulomb excitation of the Au backing. The transitions in the ground state band are observed to be in coincidence with Pu X rays as well as the 2615-keV $2^+ \rightarrow 0^+$ transition of the reaction beam partner ^{208}Pb. These coincidences help confirm the assignment of this rotational band to ^{244}Pu. Figure 2 shows the

DOI: 10.1051/epjconf/201612302003

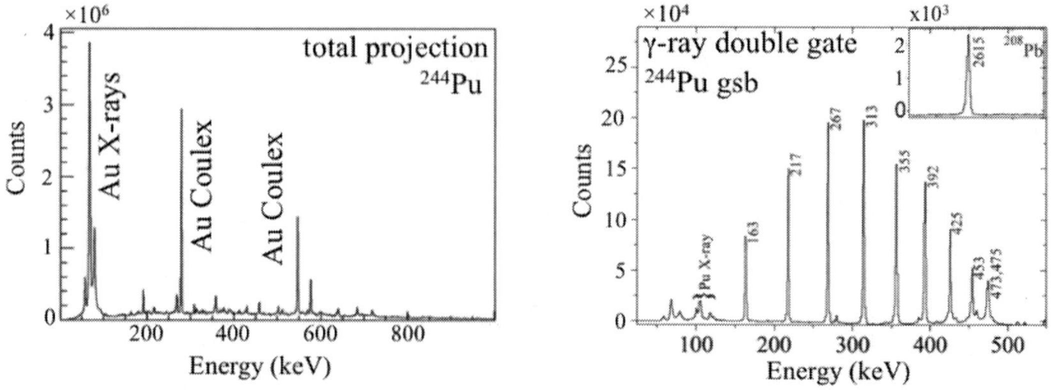

Figure 1. The total projection of a γ-γ-γ cube versus a spectrum double-gated on ground state band γ rays in ^{244}Pu (see text).

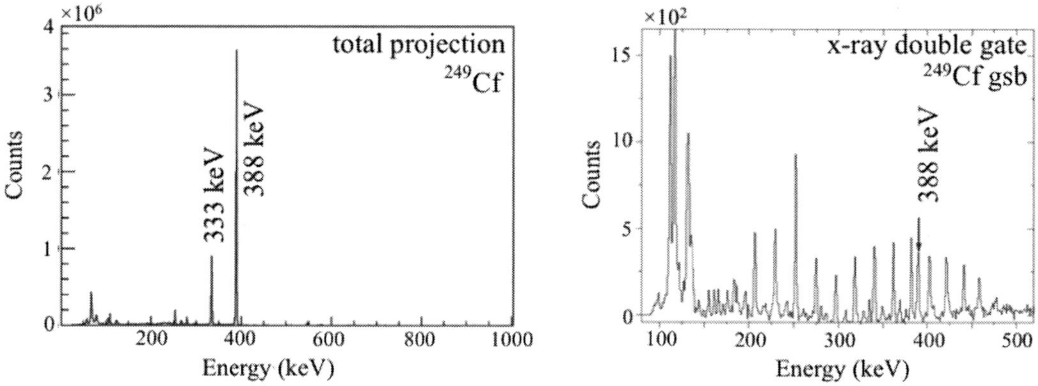

Figure 2. The total projection of a γ-γ-γ cube from the ^{249}Cf experiment versus a spectrum double-gated on Cf X rays (see text).

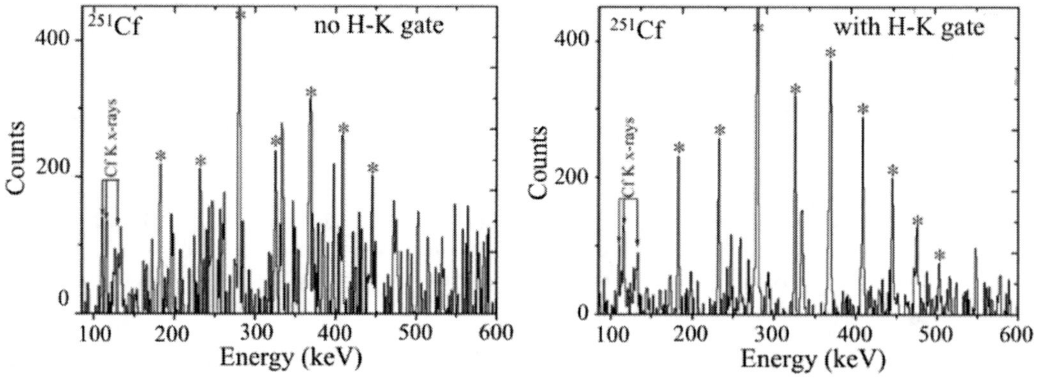

Figure 3. Comparison of ^{251}Cf spectra without and with H-K gating, demonstrating vastly improved signal-to-noise ratios (see text).

total projection of a γ-γ-γ cube for the ^{249}Cf target with an activity of $\approx$25 μCi, where the spectrum is now dominated by ^{245}Cm γ rays from the α decay of ^{249}Cf. Here, a double gate on two Cf X rays suppresses the Cm γ rays and reveals the ground state band in ^{249}Cf. Finally, Fig. 3 illustrates how gating on sum-energy (H) and gamma-ray fold (K) helps remove fission background contributions in the case of ^{251}Cf without affecting the photopeak intensities.

DOI: 10.1051/epjconf/201612302003

3 Results

Rotational excitations were populated typically to angular momenta $I > 20\hbar$ in all nuclei, with multiple band structures identified and quasiparticle alignments mapped in each nucleus [6, 7]. A couple of key experimental highlights and physics advances are briefly summarised below.

3.1 N = 151 nuclei

Band structures in odd-A nuclei help identify specific nucleon orbitals on which the rotations are built. Prior to the present work, for $N = 151$ nuclei, there was only limited spectroscopic information for ^{245}Pu [8], with only ground-state bands observed in ^{247}Cm and ^{249}Cf [9]. We have now identified two rotational band structures in ^{245}Pu built on the $\nu[734]9/2^-$ and $\nu[624]7/2^+$ configurations, as well as new rotational bands in ^{247}Cm and ^{249}Cf built on the $\nu[622]5/2^+$ configuration [6]. In each case, the configurations are deduced experimentally through measured $M1/E2$ branching ratios between the signature partners. This is the first time high-spin rotational bands built on both ground and excited state configurations, especially on orbitals other than the $j_{15/2}$ neutron orbital, have been identified and characterised in $N > 150$ systems. A long-standing unresolved issue in the $A \sim 250$ region is the fact that the expected early alignment of the $j_{15/2}$ neutrons from cranked Woods-Saxon predictions is not observed [9, 10], while the alignment of $i_{13/2}$ protons, typically predicted to occur at frequencies higher than the neutrons, has been fully mapped, e.g, in ^{244}Pu [4]. The identification of collective bands built on different configurations in the $N = 151$ isotones of Pu, Cm and Cf now allow competing alignment contributions from neutrons and protons to be disentangled using blocking arguments [6].

In ^{245}Pu and ^{247}Cm, both bands, built on different neutron orbitals, track the alignment upbends at the same frequency as their even-even ^{244}Pu and ^{246}Cm cores, respectively. Since the $j_{15/2}$ neutron alignment is blocked for the ground state $\nu[734]9/2^-$ configurations in both cases, the data suggest a common proton pair alignment. In ^{249}Cf, the $\nu[734]9/2^-$ ground state band again tracks the flat behaviour of the ^{248}Cf core, while the $\nu[622]5/2^+$ band (in which the $j_{15/2}$ neutron alignment is not blocked) shows a slight uptick and overtakes the flatter ground state band curve at the highest frequencies observed [6].

Since effects of higher order shape multipoles, especially β_6 deformation, have been reported to be significant in this mass region [11], we performed cranked Woods-Saxon calculations with and without β_6 for these $N = 151$ isotones. The predicted alignment frequencies with and without β_6 deformation are compared to experimental data in Fig. 4. The predicted proton alignment frequencies are seen to be reduced by $\approx$0.03 MeV in ^{245}Pu with the inclusion of β_6, while increasing the neutron alignment frequencies by $\approx$0.015 MeV at the same time. Without the inclusion of β_6, for all the three $N = 151$ nuclei studied, the neutrons are predicted to align earlier than the protons, contrary to experimental observations. The inclusion of this higher order shape multipole flips the order of the neutron and proton alignments in ^{245}Pu and ^{247}Cm, thus effectively bringing experiment and predictions in line for proton and neutron alignments in these nuclei (see Ref.[6] for more details).

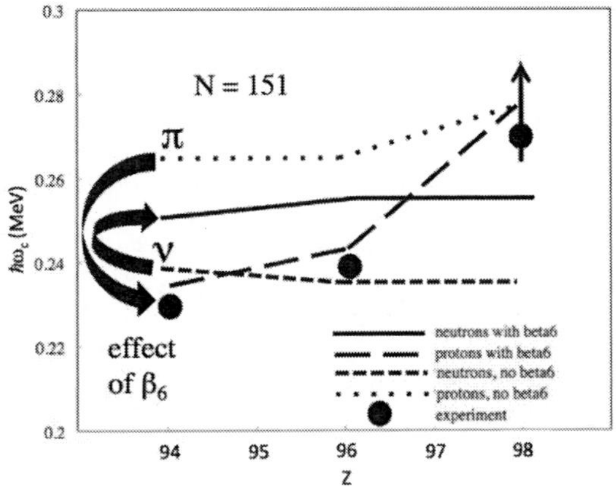

Figure 4. Effect of β_6 on neutron and proton crossing frequencies.

3.2 N > 152 nuclei

We have investigated the $N = 153$ nucleus ^{251}Cf and established the ground state band built on the $\nu[620]1/2^+$ orbital to J^π=45/2$^+$ [12]. A comparison of the alignment behaviour of ^{251}Cf with its neighbouring lighter $N = 153$ isotone ^{249}Cm [9] shows that the two align at frequencies that differ by $\approx$0.03 MeV, precluding a neutron alignment scenario and endorsing a proton alignment behaviour for both nuclei.

Finally, we have studied the $N = 154$ nucleus ^{250}Cm and extended the ground state band, observed earlier to J^π=12$^+$ [13], to J^π=24$^+$ [7]. This is the first even-even nucleus beyond the $N = 152$ shell gap that is studied to high angular momenta. A strong alignment upbend is observed at the highest frequencies, in contrast to the neighbouring lighter even-Cm isotopes. This, together with a reduced kinematic moment of inertia at low spins for the $N = 154$ isotope compared to its neighbours, points towards increased pairing correlations beyond the $N = 152$ sub-shell gap. The sharp alignment feature also provides a tantalising possibility that the crossing may involve high-j low-Ω orbitals that originate from above the $N = 184$ spherical shell gap, but have been brought down by deformation to the valence region near $N = 154$ [7, 14].

3.3 Summary and Outlook

Our experimental program with inelastic and transfer reactions with heavy beams and radioactive actinide targets using the Gammasphere array has yielded a wealth of new data in $94 < Z < 98$ nuclei. These have expanded our spectroscopic horizons in this very heavy mass region to high angular momenta, allowed us to study collective excitations built on the highest neutron orbitals,

DOI: 10.1051/epjconf/201612302003

and have served as an excellent complement to the fusion-evaporation studies of excitations in $Z > 100$ nuclei [1]. Expanding the studies to odd-Z nuclei is a natural next step for a comprehensive exploration of the $N - Z$ landscape, as very little data exists in that domain. We are currently pursuing vibrational excitations and non-axial shape degrees of freedom in this mass region from the prompt spectroscopy. Finally, K-isomer studies remain a rich hunting ground, not only to extract single-particle energies and address pairing correlations, but to explore whether and how these metastable states that arise from quantum hindrances can inform our understanding of the stability and synthesis of superheavy elements.

Acknowledgments

We gratefully acknowledge helpful discussions with our theoretical colleagues Stefan Frauendorf and Furong Xu. The trans-plutonium element production facilities at Oak Ridge National Laboratory, supported by the Office of Basic Energy Science, U.S. Department of Energy, are acknowledged for providing isotopes for the targets. This material is based upon work supported by the U.S. Department of Energy, Office of Science, Office of Nuclear Physics, under award numbers DE-FG02-94ER40848 and contract number DE-AC02-06CH11357. This research used resources of ANL's ATLAS facility, which is a DOE Office of Science user facility.

References

[1] Ch. Theisen *et al.*, Nucl. Phys. **A944**, 333 (2015).
[2] G. Henning *et al.*, Phys. Rev. Lett. **113**, 262505 (2014).
[3] G. Hackman *et al.*, Phys. Rev. C **57**, 1056(R) (1998).
[4] I. Wiedenhoever *et al.*, Phys. Rev. Lett. **83**, 2143 (1999).
[5] K. Abu Saleem, Ph.D. Thesis, Illinois Institute of Technology, (2002).
[6] S. S. Hota *et al.*, Phys. Lett. B **739**, 13 (2014).
[7] S. S. Hota, Ph.D. Thesis, University of Massachusetts Lowell, (2012).
[8] H. Makii *et al.*, Phys. Rev. C **76**, 061301(R) (2007).
[9] S. K. Tandel *et al.*, Phys. Rev. C **82**, 041301(R) (2010).
[10] K. Abu Saleem *et al.*, Phys. Rev. C **70**, 024310 (2004).
[11] A. Sobiczewski *et al.*, Phys. Rev. C **63**, 034306 (2001).
[12] Y. Qiu *et al.*, to be published.
[13] T. Ishii *et al.*, J. Phys. Soc. Japan **75**, 043201 (2006).
[14] S. S. Hota *et al.*, to be published.

EPJ Web of Conferences **123**, 02004 (2016)
Heavy Ion Accelerator Symposium 2015
DOI: 10.1051/epjconf/201612302004

Electric Monopole Transition Strengths in ^{62}Ni

L. J. Evitts[1,2,a], A. B. Garnsworthy[1], T. Kibédi[3], M. Moukaddam[1], B. Alshahrani[3,4], T. K. Eriksen[3], J. D. Holt[1], S. S. Hota[3], G. J. Lane[3], B. Q. Lee[3], B. P. McCormick[3], N. Palalani[3], M. W. Reed[3], S. R. Stroberg[1], and A. E. Stuchbery[3]

[1] *TRIUMF, 4004 Wesbrook Mall, Vancouver BC, V6T 2A3, Canada*
[2] *Department of Physics, University of Surrey, Guildford, GU2 7XH, United Kingdom*
[3] *Department of Nuclear Physics, Research School of Physics and Engineering, The Australian National University, Canberra, ACT 2601, Australia*
[4] *King Khaled University, Abha, Kingdom of Saudi Arabia*

Abstract. Excited states in ^{62}Ni were populated with a (p, p') reaction using the 14UD Pelletron accelerator at the Australian National University. Electric monopole transition strengths, $\rho^2(E0)$, were measured through simultaneous detection of the internal conversion electrons and γ rays emitted from the de-excitation of populated states, using the Super-e spectrometer coupled with a germanium detector. The strength of the 0_2^+ to 0_1^+ transition has been measured to be $77^{+23}_{-34} \times 10^{-3}$ and agrees with previously reported values. Upper limits have been placed on the 0_3^+ to 0_1^+ and 0_3^+ to 0_2^+ transitions. The measured $\rho^2(E0)$ value of the 2_2^+ to 2_1^+ transition in ^{62}Ni has been measured for the first time and found to be one of the largest $\rho^2(E0)$ values measured to date in nuclei heavier than Ca. The low-lying states of ^{62}Ni have previously been classified as one- and two-phonon vibrational states based on level energies. The measured electric quadrupole transition strengths are consistent with this interpretation. However as electric monopole transitions are forbidden between states which differ by one phonon number, the simple harmonic quadrupole vibrational picture is not sufficient to explain the large $\rho^2(E0)$ value for the 2_2^+ to 2_1^+ transition.

1 Introduction

Electric monopole ($E0$) transitions occur between nuclear states in the same nucleus that possess the same spin and parity, J^π. Due to conservation of angular momentum, single γ-ray emission is forbidden between states of $J^\pi = 0^+$ but the transition can occur via internal conversion electrons and pair production. The strength of $E0$ transitions, $\rho^2(E0)$, is a good indication of the underlying nuclear structure. For example, they provide information that can indicate a change in the nuclear shape or on the mixing between a number of states [1, 2].

For the stable even-even Ni isotopes, $\rho^2(E0)$ values for transitions between 0^+ states have been measured in ^{58,60,62}Ni [3–5] while no measurement exists for ^{64}Ni. Furthermore, $\rho^2(E0 : J^+ \to J^+)$ values for non-zero spin states have not been measured in the Ni isotopes. The previous measurements are shown in Fig. 1.

It has long been considered that ^{62}Ni is a good example of a quadrupole spherical vibrator due to the 0^+, 2^+, 4^+ triplet of states at twice the energy of the 2_1^+ state. The model is further supported by the measurement of a small quadrupole moment, $Q(2_1^+) = +0.05(12)$ b [6], and the collective B($E2$) values [7, 8]. However, the spherical vibrator model has faced challenges in recent years e.g., Garrett *et al.* have suggested the breakdown of vibrational excitations in the Cd isotopes [9, 10]. In the Cd case it seems that mixing with multi-quasiparticle intruder configurations are likely to play a significant role in the development of apparent collectivity in the ground-state structure.

Chakraborty *et al.* [8] measured the B($E2\downarrow$: $2_1^+ \to 0_1^+$) value in ^{62}Ni to be 12.1(4) W.u., which is in excellent agreement with a theoretical value of 12 W.u. from the vibrator model. However, they also measured the B($E2\downarrow$: $2_2^+ \to 2_1^+$) value for the transition between the suspected two-phonon and one-phonon states to be $12.5^{+3.8}_{-3.2}$ W.u., which is 3σ away from the expected theoretical value based on a spherical quadrupole vibrator of 24 W.u.. A detailed examination of the electric monopole transition strengths is expected to help clarify our understanding of the underlying nature of these excitations.

Various models can be employed to make predictions of electric monopole transition strengths. In the spherical vibrator model the $2_2^+ \to 2_1^+$ transition is forbidden via $E0$ decay as it would be in violation of the $\Delta N_{phonon} = \pm 0, 2$ selection rule. An estimation of $\rho^2(E0 : 0_2^+ \to 0_1^+) \times 10^3 = 27(1)$ can be made from

$$\rho^2(E0 : 0_2^+ \to 0_1^+) = \frac{2}{5}\frac{B(E2\uparrow: 0_1^+ \to 2_1^+)^2}{(3/4\pi)^2 Z^2 r_0^8 A^{8/3}}, \quad (1)$$

[a] e-mail: evitts@triumf.ca

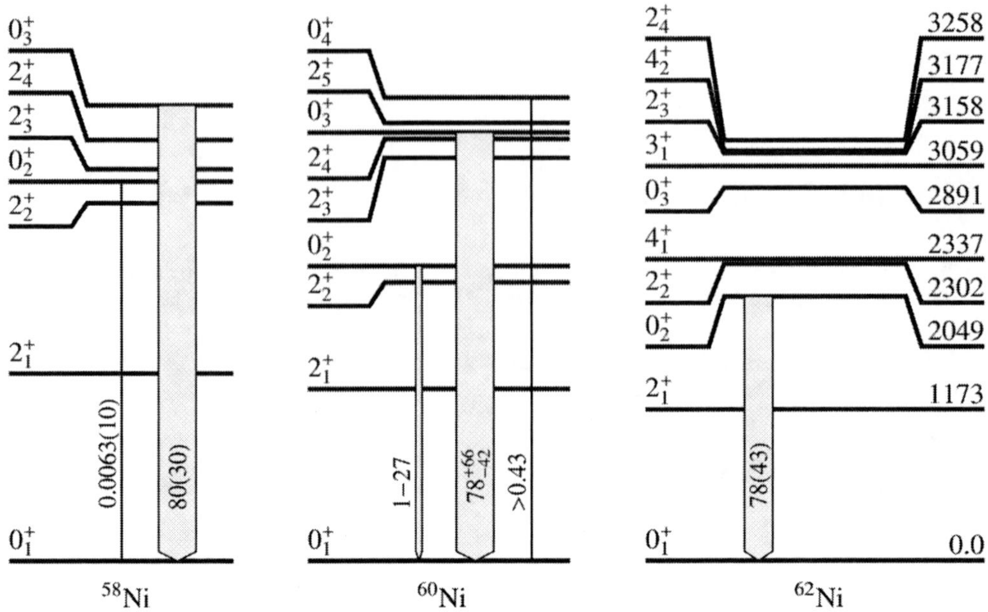

Figure 1. The low-lying states in ^{58,60,62}Ni. Only the 0^+ and 2^+ states are shown for ^{58,60}Ni. The previously measured $E0$ transitions, along with their associated strength, are shown by the arrows [5]. Note that only $0^+ \rightarrow 0^+$ transitions have been measured previously.

where Z is the atomic number, A is the mass number, r_0 is 1.3 fm and $B(E2)$ has units of e^2 fm^4 [11]. If ^{62}Ni is a rigid rotor and the first 0^+ state is assumed to be a beta vibration, then the $\rho^2(E0) \times 10^3$ values between the beta band and the ground-state band (independent of angular momentum) are estimated to be 261(7) from

$$\rho^2(E0) = \frac{9}{8\pi^2} Z^2 \beta_0^4 \frac{E(2_1^+)}{E(0_\beta^+)}$$
$$= \frac{B(E2: 0_1^+ \rightarrow 2_\beta^+) \uparrow 4\beta_0^2}{e^2 r_0^4 A^{4/3}}, \quad (2)$$

where E is the energy, β refers to the beta band, e is the charge of the electron and β_0 is the quadrupole deformation [11].

2 Experiment

The 14UD Heavy Ion accelerator at the Australian National University was used to bombard an isotropically enriched ^{62}Ni target inside the Super-e electron spectrometer [12, 13] with a proton beam. A target and tuning aperture are available on a common actuator. The 5-mm diameter tuning aperture is coated with scintillating chromium-doped aluminium oxide. During beam tuning a light-sensitive camera is used to visualize the beam spot size and position. The second target position is the reaction target and it is tilted at 45 degrees to the beam axis, as shown in Fig. 2, such that the electrons and positrons are emitted towards the electron detectors through a smaller effective target thickness and therefore experience less energy straggling.

The spectrometer operates using a solenoid magnet, with the central axis arranged perpendicular to the beam direction. Electrons from the target are guided to an array of six 9-mm thick lithium-drifted Silicon [Si(Li)] detectors which are located 350 mm away from the target and cooled with liquid nitrogen. The strength of the magnetic field of the solenoid is adjusted in order to select a momentum window for the transmission of a particular electron or positron energy. Two baffles and a diaphragm are located between the target and the detector, with the purpose of shielding the detectors from direct illumination of γ radiation from the target, and to prevent the transmission of low-energy electrons along the central axis. Electrons and positrons cross the central axis twice before reaching the detectors completing 2.5 helical orbits, as shown in the schematic of the spectrometer in Fig. 2. During normal operation, the magnetic field is swept back and forth between the minimum and maximum magnetic field values, covering the range of expected transition energies.

The relative efficiency of the HPGe detector was determined by the use of a ^{56}Co source. The relative efficiency of the Super-e spectrometer was determined by populating states in ^{54}Fe through inelastic proton scattering at 6.9 MeV and measuring the internal conversion coefficients of the pure $E2$ transitions that de-populate the excited states. Excited states in ^{62}Ni were populated through inelastic proton scattering of a 500-nA, 5-MeV proton beam incident on a 1.2-mg/cm^2 thick self-supporting ^{62}Ni target for a period of 24.5 hours.

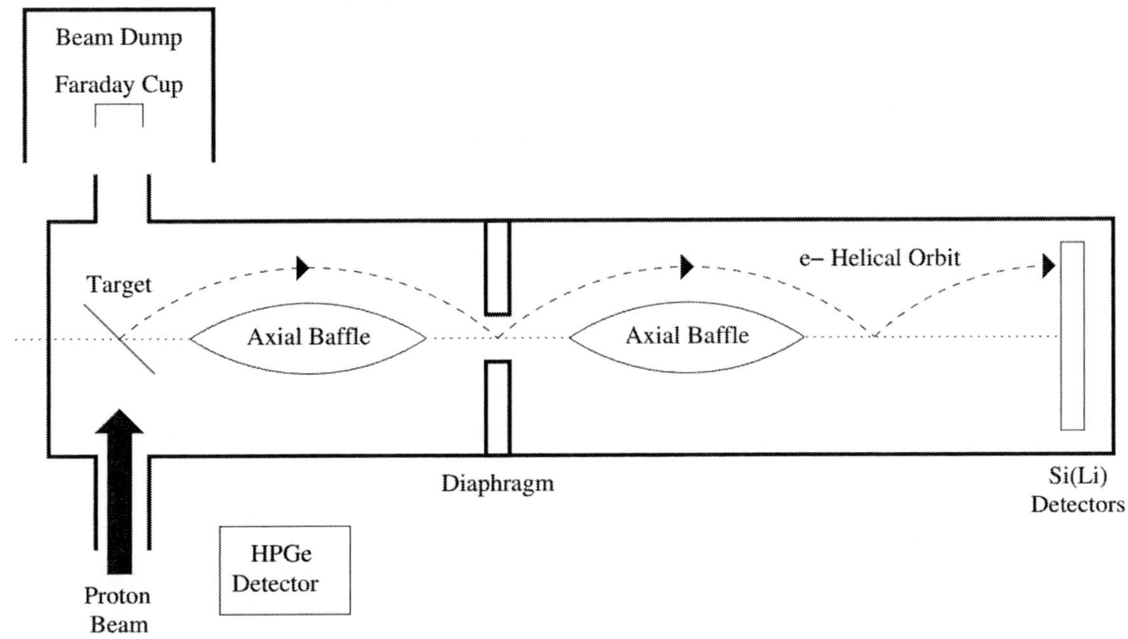

Figure 2. A schematic of the Super-e spectrometer. A proton beam is incident on a tilted target; the electrons are emitted outside of the rear and transported by the means of a solenoid magnet around two axial baffles and through a diaphragm onto an array of Si(Li) detectors.

3 Results

The spectrometer has a finite momentum acceptance window for each magnetic field setting. The signal-to-background ratio of the electron spectra can be improved if a software condition is imposed on the fraction of B to $B(\rho)$ in the event by event data. The momentum-window gated electron spectrum, along with the Ge singles, is shown in Fig. 3. The observed transitions in ^{62}Ni are shown by the dashed lines, the other transitions in the Ge singles are from the ^{63}Cu (p, γ) reaction. An excitation function was performed for beam energies between 5 and 6.5 MeV. A beam energy of 5 MeV was chosen in order to maximize the cross section for populating excited states in ^{62}Ni whilst keeping the cross section of the ^{63}Cu (p, γ) and ^{62}Cu (p, nγ) channels to a manageable level.

In order to validate the method of determining $\rho^2(E0)$ values and our understanding of the relative efficiencies, a q^2 value (i.e. the ratio of $E0$ to $E2$ electrons) of 0.088(5) was measured for the $0_2^+ \rightarrow 0_1^+$ transition in ^{62}Ni. The uncertainty in q^2 is calculated from the statistical uncertainty and systematic error in the efficiency of the Super-e detector. This value agrees within 1σ of the previously measured value of 0.084(11) [3, 5]. The mean lifetime, τ, of the 0_2^+ state at 2048.68 keV has been measured in a ^{59}Co(α, pγ) reaction as $1.1^{+1.1}_{-0.4}$ ps [15], and in a ^{62}Ni(n, n'γ) reaction as $2.6^{+2.7}_{-0.9}$ ps [8]. Thus, the weighted average of the two data points was calculated to be $1.3^{+1.0}_{-0.3}$ ps. The $\rho^2(E0)$ value was determined to be $77^{+23}_{-34} \times 10^{-3}$ by using the mean-lifetime and

$$\rho^2(E0) = q_K^2 \times \frac{\alpha_K(E2)}{\Omega_K(E0)} \times W_\gamma(E2), \quad (3)$$

where Ω_K is the electronic factor and α_K and W_γ are the theoretical internal conversion coefficient [16] and partial decay constant of the $0_2^+ \rightarrow 2_1^+$ transition [5]. By utilizing similar methods, the $\rho^2(E0)$ value of the $2_2^+ \rightarrow 2_1^+$ was found to be roughly three times larger than the 0_2^+ to 0_1^+ transition. It was also possible in this experiment to place upper limits for the 0_3^+ to 0_1^+ and 0_3^+ to 0_2^+ transitions.

4 Summary

The $\rho^2(E0)$ values in ^{62}Ni for transitions between $J_i^\pi = J_f^\pi$ states were measured using the Super-e spectrometer at the Australian National University. The strength of the 0_2^+ to 0_1^+ transition was re-measured to be $77^{+23}_{-34} \times 10^{-3}$ and agrees with previously reported values. Upper limits have been placed on the 0_3^+ to 0_1^+ and 0_3^+ to 0_2^+ transitions. The measured $\rho^2(E0)$ value of the 2_2^+ to 2_1^+ transition in ^{62}Ni has been measured for the first time and found to be one of the largest $\rho^2(E0)$ values measured to date in nuclei heavier than Ca. This large value for the 2_2^+ to 2_1^+ transition can not be sufficiently explained by the simple harmonic quadrupole vibrational picture. The full results from this study will be presented in a future publication [17]. Similar investigations of $E0$ transition strengths between $J_i^\pi = J_f^\pi$ states in the other stable nickel isotopes are underway using the same experimental setup at the ANU.

 DOI: 10.1051/epjconf/201612302004

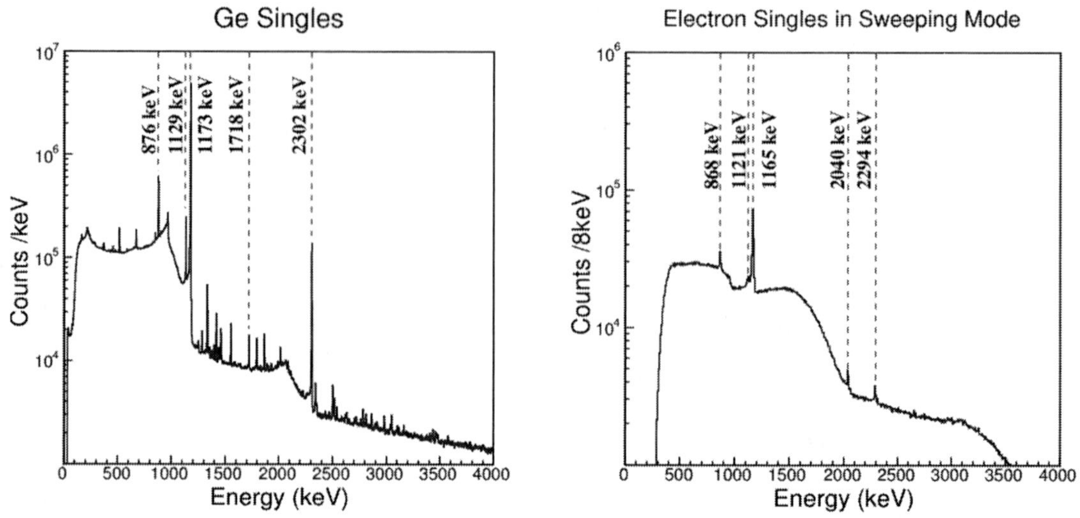

Figure 3. The energy spectra for (left) the Ge singles and (right) the electron singles in sweeping mode of the spectrometer, over the whole experimental run. The labelled peaks are the transitions observed from the de-excitating states in ^{62}Ni. In the electron spectrum, the K conversion electron peaks are highlighted, where the K-shell electron binding energy in Ni is 8.3 keV [14]

References

[1] K. Heyde and J. Wood, Rev. Mod. Phys. **83**, 1467 (2011).
[2] G. Ilie and R. F. Casten, Phys. Rev. C **84**, 064320 (2011).
[3] A. Passoja *et al.*, Nucl. Phys. A **363**, 399 (1981).
[4] E. K. Warburton and D. E. Alburger, Phys. Lett. B **36**, 38 (1971).
[5] T. Kibédi and R. H. Spear, Atomic Data and Nuclear Data Tables **89**, 77 (2005).
[6] J. M. G. Gómez, Phys. Rev. C **6**, 149 (1972).
[7] N. J. Stone, At. Data Nucl. Data Tables **90**, 75 (2005).
[8] A. Chakraborty *et al.*, Phys. Rev. C **83**, 034316 (2011).
[9] P. E. Garrett *et al.*, Phys. Rev. C **78**, 044307 (2008).
[10] P. E. Garrett and J. L. Wood, J. Phys. G **37**, 064028 (2010).
[11] J. L. Wood *et al.*, Nucl. Phys. A **651**, 323-368 (1999).
[12] T. Kibédi *et al.*, Eur. Phys. J. Web Conf. **35** 06001 (2012).
[13] T. Kibédi *et al.*, in proceedings of "XII. International Symposium on Nuclei in the Cosmos", PoS(NIC XII)203 (2012).
[14] J. A. Bearden, Rev. Mod. Phys. **39**, 78 (1967).
[15] D. L. Kennedy *et al.*, Nucl. Phys. A **308**, 14 (1978).
[16] T. Kibédi *et al.*, Nucl. Inst. and Meth. A **589**, 202 (2008).
[17] L. J. Evitts *et al.*, *to be published* (2016).

EPJ Web of Conferences **123**, 02005 (2016) DOI: 10.1051/epjconf/201612302005
Heavy Ion Accelerator Symposium 2015

Conversion electrons from high-statistics β-decay measurements with the 8π spectrometer at TRIUMF-ISAC

P. E. Garrett[1,a], B. Jigmeddorj[1], A. J. Radich[1], C. Andreoiu[2], G. C. Ball[3], J. C. Bangay[1], L. Bianco[1,b], V. Bildstein[1], S. Chagnon-Lessard[1,c], D. S. Cross[2], G. A. Demand[1], A. Diaz Varela[1], R. Dunlop[1], P. Finlay[1,d], A. B. Garnsworthy[3], K. L. Green[1], G. Hackman[3], B. Hadinia[1], K. G. Leach[1,e], J. Michetti-Wilson[1], J. N. Orce[3,4], M. M. Rajabali[3,f], E. T. Rand[1], K. Starosta[2], C. Sumithrarachchi[1,g], C. E. Svensson[1], S. Triambak[3,4,5], Z. M. Wang[2,3], S. J. Williams[3,h], J. L. Wood[6], J. Wong[1], S. W. Yates[7], and E. F. Zganjar[8]

[1] *Department of Physics, University of Guelph, Guelph, ON N1G2W1 Canada*
[2] *Department of Chemistry, Simon Fraser University, Burnaby, BC V5A1S6 Canada*
[3] *TRIUMF, 4004 Wesbrook Mall, Vancouver, BC V6T2A3 Canada*
[4] *Department of Physics, University of the Western Cape, P/B X17, Bellville, ZA-7535 South Africa*
[5] *iThemba LABS, P.O. Box 722, Somerset West 7129, South Africa*
[6] *School of Physics, Georgia Institute of Technology, Atlanta, Georgia 30332, USA*
[7] *Departments of Chemistry and Physics & Astronomy, University of Kentucky, Lexington, Kentucky 40506, USA*
[8] *Department of Physics and Astronomy, Louisiana State University, Baton Rouge, Louisiana 70808, USA*

Abstract. The 8π spectrometer, located at TRIUMF-ISAC, was the world's most powerful spectrometer dedicated to β-decay studies until its decommissioning in early 2014 for replacement with the GRIFFIN array. An integral part of the 8π spectrometer was the Pentagonal Array for Conversion Electron Spectroscopy (PACES) consisting of 5 Si(Li) detectors used for charged-particle detection. PACES enabled both $\gamma - e^-$ and $e^- - e^-$ coincidence measurements, which were crucial for increasing the sensitivity for discrete e^- lines in the presence of large backgrounds. Examples from a ^{124}Cs decay experiment, where the data were vital for the expansion of the ^{124}Cs^m decay scheme, are shown. With sufficient statistics, measurements of conversion coefficients can be used to extract the $E0$ components of $J^\pi \rightarrow J^\pi$ transitions for $J \neq 0$, which is demonstrated for data obtained in ^{110}In$\rightarrow^{110}$Cd decay. With knowledge of the shapes of the states involved, as obtained, for example, from the use of Kumar-Cline shape invariants, the mixing of the states can be extracted.

1 Introduction

Prior to its decommissioning, the 8π spectrometer, located at the TRIUMF-ISAC radioactive beam facility in Vancouver, Canada, was the world's most powerful HPGe array routinely used for β-decay studies. This status was achieved because the spectrometer and its auxiliary detector systems were dedicated to β-decay studies only, allowing it to be optimized for that singular purpose. The 20 Compton-suppressed HPGe γ-ray detectors were configured in a truncated icosahedral geometry and achieved $\approx$ 1% photopeak efficiency at the energies of the ^{60}Co γ rays. Surrounding the beam-implantation point onto a FeO or Al coated mylar tape of the Moving Tape Collector at the center of a vacuum chamber were up to 20 plastic scintillator detectors of the SCintillating Electron-Positron Tagging Array (SCEPTAR). A different configuration involved the removal of the upstream hemisphere of SCEPTAR and its replacement with the Pentagonal Array for Conversion Electron Spectroscopy (PACES). Finally, the downstream SCEPTAR hemisphere could be removed and replaced with a single fast-plastic scintillator. These detector systems were coupled to a high-throughput and high-precision data acquisition system capable of triggering in excess of 30 kHz that enabled the collection of high-statistics data sets in a reasonable amount of beam time, but also enabled the half life and branching ratio measurements to a level of precision of better than 0.1% required for studies of Fermi superallowed β^+ emitters. Additional details of the spectrometer can be found in Refs. [1–3].

[a] e-mail: pgarrett@physics.uoguelph.ca
[b] present address: DESY Photon Science, Notkestrasse 85 D-22607 Hamburg, Germany
[c] Present address: Department of Physics, University of Ottawa, 150 Louis-Pasteur, Ottawa, ON K1N 6N5, Canada
[d] Present address: Instituut voor Kern-en Stralingsfysica, K.U. Leuven, Celestijnenlaan 200D, B-3001 Leuven, Belgium
[e] Present address: Department of Physics, Colorado School of Mines, Golden, Colorado 80401, USA
[f] Present address: Department of Physics, Tennessee Technological University, Cookeville, Tennessee 38505, USA
[g] Present address: National Superconducting Cyclotron Laboratory, Michigan State University, 640 South Shaw Lane, East Lansing, Michigan 48824, USA
[h] Present address: Diamond Light Source Ltd., Diamond House, Harwell Science and Innovation Campus, Didcot, Oxfordshire OX11 0DE, United Kingdom

DOI: 10.1051/epjconf/201612302005

The 8π spectrometer was decommissioned in the early part of 2014 and replaced with the much more powerful GRIFFIN array [1, 4]. Consisting of 16 large-volume HPGe clover detectors, GRIFFIN utilizes all of the existing auxiliary detectors and in addition can be coupled with the DESCANT neutron detector array [5] for studies of β-delayed neutron emitters.

One of the main goals of the science program carried out with the 8π spectrometer was the investigation of collectivity in nuclei, especially those located on or near the valley of stability. For these studies, high sensitivity to weak, low-energy γ-ray branches from highly-excited states is necessary. Studies to date have included, for example, the decay of ^{110}In [6–9], ^{112}In/^{112}Ag [10, 11], and ^{122,124,126}Cs [12, 13] into ^{110}Cd, ^{112}Cd, and ^{122,124,126}Xe, respectively. The 8π spectrometer proved itself a valuable tool for these studies since, for the above species, the high intensity of the radioactive beams delivered by TRIUMF more than compensated for the low absolute efficiency (by modern standards) for γ-ray detection.

In addition to the detection and measurement of weak γ-ray decay branches, in many of the experiments the 8π spectrometer was augmented with PACES for the purpose of conversion-electron detection and to seek possible $E0$ branches. As is well known, the measurement of $E0$ strengths, in the form of $\rho^2(E0)$ values, provides crucial information concerning the mixing of states with different intrinsic shapes. Such measurements, however, prove to be especially challenging requiring the ability to extract $E0$ branches in the presence of large backgrounds from β particles and Compton scattering of γ rays. Further, high statistics for $J^\pi \rightarrow J^\pi$ transitions with $J^\pi \neq 0$ must be acquired since the $M1/E2$ contribution to the electron intensity must be subtracted.

2 E0 conversion electron studies

$E0$ transitions are sensitive to the changes in the nuclear charge-squared radii since the $E0$ operator is [14]

$$\hat{M}(E0) = \sum_i e_i r_i^2, \qquad (1)$$

where the sum extends over the A bodies in the nucleus with their charges e_i and radial position r_i. The usual quantity quoted when referring to an $E0$ transition is the $\rho^2(E0)$ value, defined via [14]

$$\Gamma(E0) = \frac{1}{\tau(E0)} = \rho^2(E0) \sum_j \Omega_j(Z, \Delta E), \qquad (2)$$

where $\Gamma(E0)$ is the partial width for the decay, $\tau(E0)$ the partial lifetime, and $\Omega(Z, E_e)$ the electronic factor that depends on the atomic number Z and the energy of the transition ΔE. The quantity $\rho(E0)$ is defined by, with $R = 1.2A^{1/3}$ fm,

$$\rho(E0) = \frac{1}{eR^2} \left\langle I_f \left| M(E0) \right| I_i \right\rangle, \qquad (3)$$

and carries all of the nuclear structure information. A successful measurement of the $\rho^2(E0)$ value thus requires a measurement of the branching fraction of the total transition intensity that proceeds through the $E0$ decay, i.e.,

$$BR(E0) = \frac{I(E0)}{\sum_i I_i(\gamma) + I_i(e^-)}, \qquad (4)$$

where the summation extends over all possible branches (neglecting very small possible contributions from two-photon decay, internal pair formation, etc.) and also a measurement of the level lifetime τ. The number of $\rho^2(E0)$ values known is still rather limited [14, 15].

The $\rho^2(E0)$ values can be expressed in terms of the collective variables from the Bohr model. The operator is [16]

$$\hat{M}(E0) = \frac{3Z}{4\pi}\left(\frac{4\pi}{5} + \beta^2 + \frac{5\sqrt{5}}{21\sqrt{\pi}}\beta^3 \cos\gamma\right), \qquad (5)$$

so that in a two-level mixing solution with quadrupole deformation parameters (β_1, γ_1) and (β_2, γ_2) the $E0$ strength is given by [14, 16]

$$\rho^2(E0) = \left(\frac{3Z}{4\pi}\right)^2 a^2\left(1 - a^2\right)\left[\left(\beta_1^2 - \beta_2^2\right) + \frac{5\sqrt{5}}{21\sqrt{\pi}}\left(\beta_1^3 \cos\gamma_1 - \beta_2^3 \cos\gamma_2\right)\right]^2. \qquad (6)$$

The first term, proportional to the differences in the squares of the quadrupole deformation parameters β, is the usual expression that one sees for the $\rho^2(E0)$ values. The second term shows the leading order term involving the shape parameter γ; since it enters as the cube of the deformation β, its contribution is usually small. What is clear from Eqn. 6 is if information on the shape parameters β are known, the mixing of states can be determined.

3 The PACES Apparatus

PACES, shown in the photograph in Fig. 1, is installed internally in the vacuum chamber that is surrounded by the HPGe detectors of the 8π spectrometer. The 5 Si(Li) detectors have a 5 mm thickness, with a surface area of ≈ 200 mm^2, and are located at a distance of 3 cm from the beam deposition position. The Si(Li) detectors are in thermal contact with an Al plate that is cooled by an annular Cu cold finger of 0.75 m length. The annular cold finger is placed inside the beam line, with the annular opening allowing the beam to pass through. The Si(Li) detectors have a typical resolution of 2.5 keV at 1 MeV. The available active area of PACES covers $\approx 7\%$ of 4π solid angle, making it an efficient device for coincidence studies. Further details can be found in Refs. [1, 8].

4 Conversion electron spectroscopy with PACES

Inclusion of conversion-electron data may reveal not only $E0$ transitions, indicative of shape coexistence effects, but also provides multipolarity information. In studies of nuclei with a large number of levels at low-excitation energy,

 DOI: 10.1051/epjconf/201612302005

Figure 1. Photograph showing the detail of the PACES array of Si(Li) detectors located on the upstream side of the vacuum chamber of the 8π spectrometer. Details of the positioning are listed in Ref. [8].

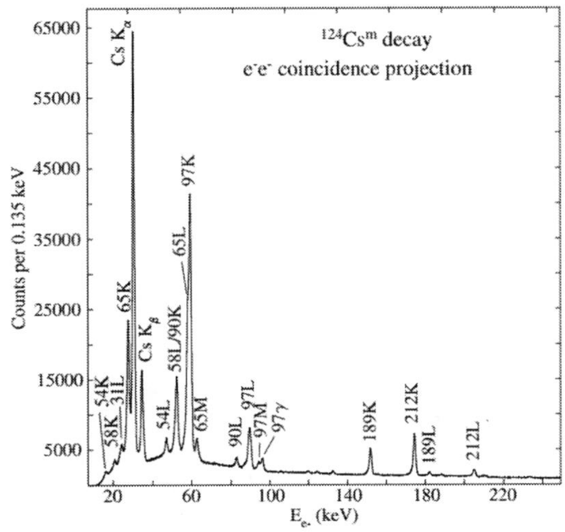

Figure 3. Projection of the e^-– e^- coincidence matrix obtained during a measurement of the decay of $^{124}Cs^m$.

as is often encountered in odd-odd nuclei or in the actinide region, the ability to detect low-energy conversion electrons and γ rays is vital in the construction of accurate decay schemes.

As mentioned above, a limitation on the measurement of $E0$ transitions, for $J \neq 0$ states, is that both the background level and the contribution of the $M1/E2$ electrons to the observed intensity of the transition must be subtracted. This requires much higher statistics than a measurement of the $0^+ \rightarrow 0^+$ transitions. Figure 2 displays a spectrum observed with the Si detectors of PACES in a recent measurement of the ^{124}Cs β^+/EC decay [13]. The sudden increases in the level of the background observed at 210 keV and 340 keV are Compton edges from intense 354-keV γ rays and the 511-keV annihilation radiation.

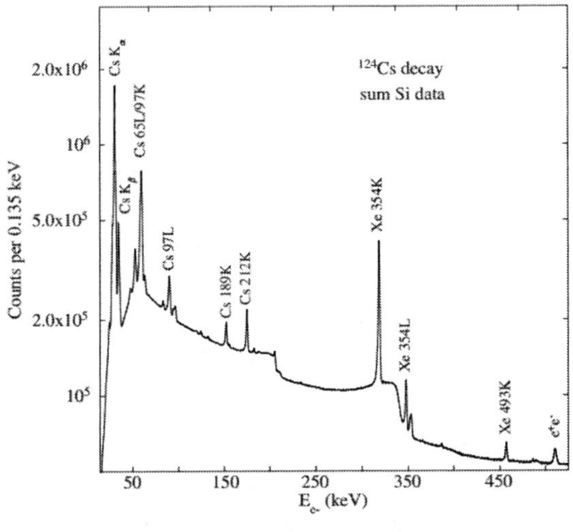

Figure 2. Example of a spectrum of conversion electrons obtained with the Si detectors of the 8π spectrometer in the β^+/EC decay of ^{124}Cs. The background is due to β^+-particle interactions in the detector, as well as Compton-scattering of γ rays. The main discrete lines observed are from Cs X-rays, transitions from the internal decay of the ^{124}Cs 7^+ isomeric state, and the most intense transitions in ^{124}Xe.

The fact that PACES consists of multiple, independent Si detectors enables $e^- - e^-$ coincidence spectroscopy. During a portion of a measurement to study the decay of ^{124}Cs, the observation of the decay of the $t_{1/2}$ = 6.3-s (7^+) ^{124}Cs isomeric state was enhanced relative to the $t_{1/2}$ = 30.8-s 1^+ ground state by employing a beam-on and tape movement cycling consisting of 1 s of beam implantation followed by 12 s of decay. This cycling minimized the build-up of the longer-lived ground state on the collection tape. Figure 3 shows the projection of the $e^- - e^-$ coincidence matrix collected during this portion of the experiment. Compared with data presented in Fig. 2, which was obtained using a beam/tape cycling appropriate for the long-lived ^{124}Cs ground state, the background in Fig. 3 is much lower due to the lower relative intensities of the high-energy γ rays in ^{124}Xe and the β^+ particles. Figure 4 displays $e^- - e^-$ coincidence spectra from gates with the 65-keV M conversion electrons (top panel) and 97-keV L electrons (bottom panel) from the ^{124}Cs isomer decay. Of particular note is the presence in the bottom panel of the 25-keV peak from the 31-keV L conversion electron, clearly resolved from the Cs K_α X-ray; the 31-keV transition was previously unobserved. Using data such as this, the decay scheme of the ^{124}Cs isomer was significantly modified and is shown in Fig. 5.

 DOI: 10.1051/epjconf/201612302005

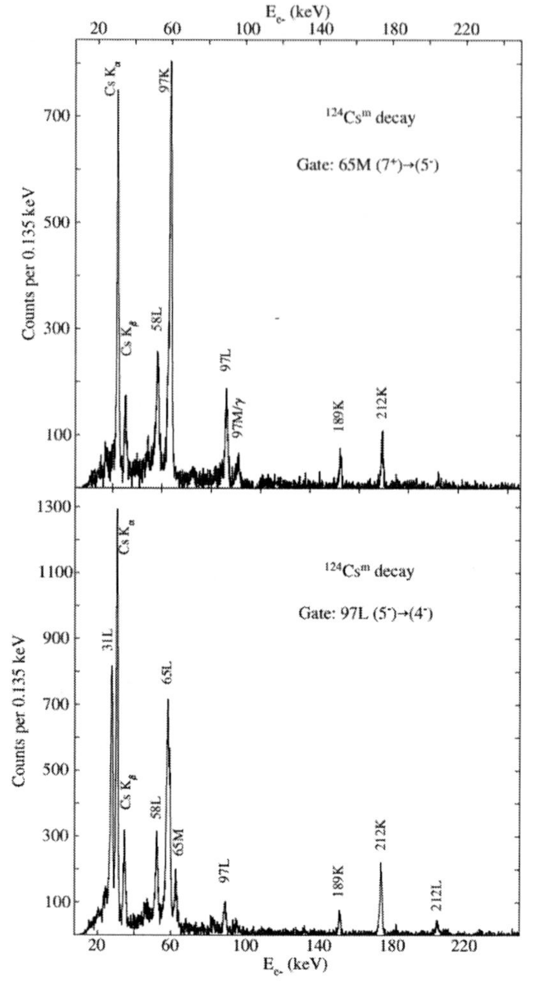

Figure 4. Examples of e^-– e^- coincidence spectra from the decay of ^{124}Cs^m. The top panel displays the coincidence spectrum with a gate taken on the 65-keV M conversion electron peak, whereas the bottom panel displays the coincidence spectrum with the 97-keV L conversion electron peak. The 97-keV L gate includes a small contribution from the 90-keV γ-ray photopeak.

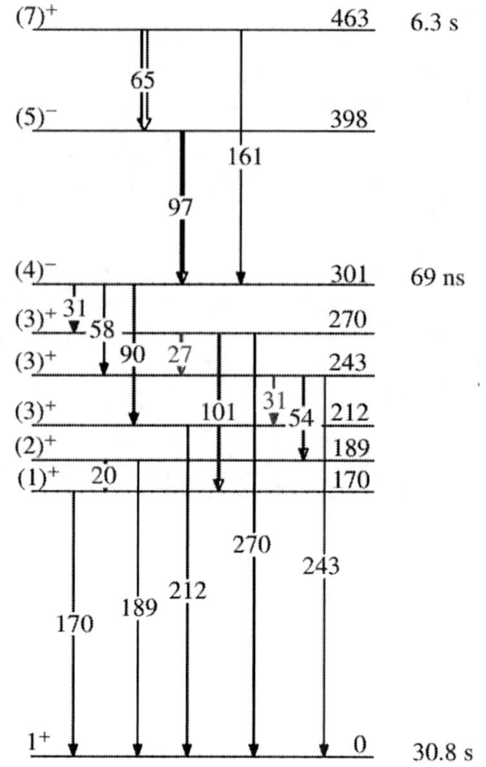

Figure 5. Level scheme deduced for the isomeric decay of ^{124}Cs^m. Levels are labelled with their energies in keV, and their J^π values. The transitions are labelled with their energies in keV. Arrows in blue are previously unobserved transitions in the decay of the isomer, with those in red new transitions.

5 E0 transitions in ^{110}Cd

A high-statistics data set for the decay of ^{110}In→^{110}Cd was collected with the dual goals of measuring weak, low-energy γ-ray decay branches from highly-excited states, and observing additional $E0$ branches. The high-intensity ion beam of 1.2×10^7 ions/s of ^{110}In in the 7^+ ground state with a 4.9-hr half life, and 1.7×10^6 ions/s of ^{110}In in the 2^+ isomeric state with a 69-min half life was delivered to the center of the 8π spectrometer.

The internal conversion coefficients were extracted from the $\gamma - e^-$ and $\gamma - \gamma$ coincidence data. The γ-ray coincidence gates were taken from above or below the decay of interest. When gating from below, the number of $\gamma - e^-$ coincident counts measured can be expressed as

$$N_{\gamma e^-} = K I^f_{e^-} \epsilon^f_{e^-} BR^d_\gamma \epsilon^d_\gamma \epsilon_c \eta(\theta_{\gamma e^-}), \tag{7}$$

where K is an overall normalization factor, I_{e^-} is the conversion electron intensity (f is the feeding and d the draining transition), ϵ_{e^-} is the detector efficiency for the conversion electron, ϵ_γ is the detector efficiency for the γ ray, BR_γ is the γ-ray branching ratio, ϵ_c is the coincidence efficiency and $\eta(\theta_{\gamma e^-})$ is an angular correlation factor. The number of $\gamma - \gamma$ coincident counts can be similarly written as

$$N_{\gamma\gamma} = K I^f_\gamma \epsilon^f_\gamma BR^d_\gamma \epsilon^d_\gamma \epsilon_c \eta(\theta_{\gamma\gamma}), \tag{8}$$

where I_γ is the γ-ray intensity. Rearranging for the intensity of γ rays and the intensity of conversion electrons from Eqns. 7 and 8, the internal conversion coefficient, in accordance with its definition, can be written as

$$\alpha = \frac{I^f_{e^-}}{I^f_\gamma} = \frac{N_{\gamma e^-}}{N_{\gamma\gamma}} \frac{\epsilon^f_\gamma}{\epsilon^f_{e^-}}. \tag{9}$$

An internal efficiency calibration was adopted (see Fig. 4 in Ref. [9]), using the conversion coefficients for well-

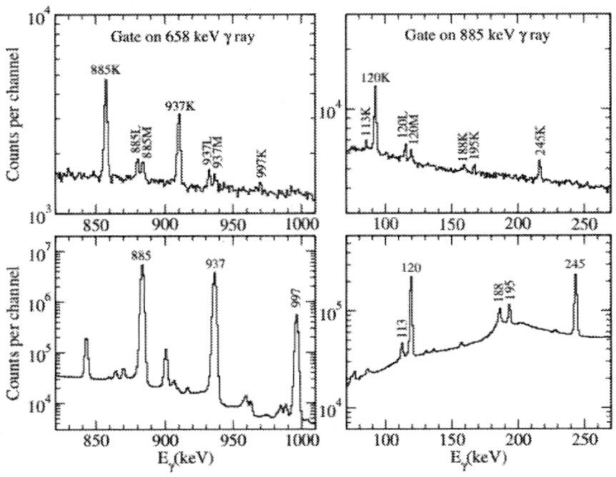

Figure 6. Portions of the conversion electron spectra in coincidence with the 658-keV $2_1^+ \rightarrow 0_{gs}^+$ (left) and 885-keV $4_1^+ \rightarrow 2_1^+$ (right) γ rays (top), and corresponding portions of the γ-ray coincidence spectra (bottom).

known $E2$ and $E1$ transitions or those determined in the present work to be essentially pure $M1$ or $E2$ transitions.

Shown in the top panel of Fig. 6 are examples of partial e^- spectra obtained by placing coincidence gates on the 658- (left) and 885-keV (right) γ rays. The bottom panel displays the corresponding γ-ray coincidence spectra with the same γ-ray coincidence gates applied. Fitting of the corresponding electron and γ-ray peaks provides the number of coincidence counts used in Eqn. 9.

Figure 7 shows portions of the e^- spectra with coincidence gates taken on the 245- and 626-keV γ rays. These particular spectra display the K-conversion electron lines from the 708-keV $4_3^+ \rightarrow 4_1^+$ transition and the 397-keV $6_2^+ \rightarrow 6_1^+$ transition that connect the deformed intruder band, based on $\pi(2p-4h)$ configuration, to the less deformed "normal" states. The extracted $\rho^2(E0)$ values for the $E0$ transitions, which includes the new value for the 708-keV $4_3^+ \rightarrow 4_1^+$ transition of $\rho^2(E0) = 106^{+98}_{-91} \times 10^{-3}$, are shown in Fig. 8.

6 Implications of ρ^2(E0) values

As shown in Eqn. 6, the determination of experimental $\rho^2(E0)$ values enables the admixtures between states to be determined provided that the underlying intrinsic shapes are known. In many cases this is not possible since the shapes have typically not been determined for excited states. However, when Coulomb excitation has been performed with sufficient sensitivity and statistics such that the Kumar-Cline [17, 18] shape invariants can be constructed, the shapes of the states can be extracted in a model-independent way. Such a case is ^{114}Cd [19].

The Coulomb excitation experiment performed by Fahlander *et al.* [19] provided a sufficient number of matrix elements that enables the shape invariant Q^2 to be de-

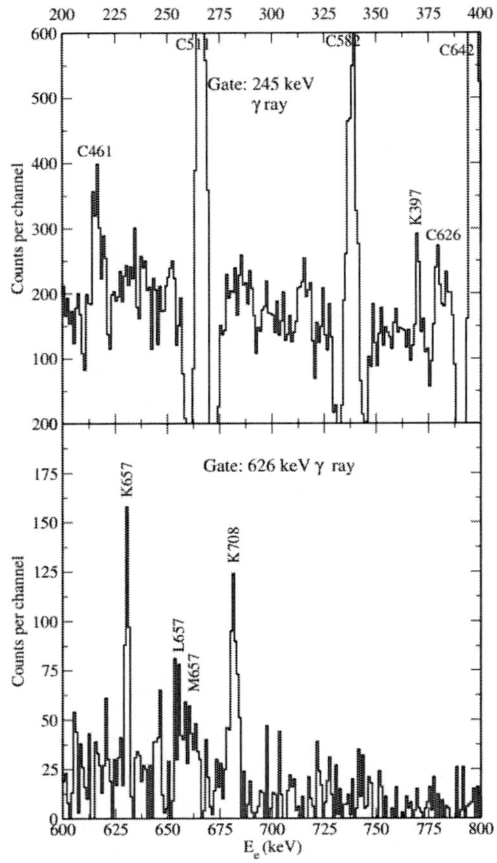

Figure 7. Portions of the spectra of the conversion electrons in coincidence with the 245- (top) and 626-keV (bottom) γ rays. The coincidence with the 245-keV γ ray shows the 397-keV K-conversion electron peak, which is weak, and some much stronger peaks that result from Compton scattering of the 511-keV annihilation radiation and the 582- and 642-keV γ rays. The coincidence spectrum with the 626-keV γ ray shows the 708-keV K-conversion electron peak.

termined from

$$\frac{1}{\sqrt{5}}Q^2 = \frac{(-1)^{2I_i}}{\sqrt{2I_i+1}} \sum_j \langle I_i \| \mathcal{M}(E2) \| I_j \rangle \times \langle I_j \| \mathcal{M}(E2) \| I_i \rangle \begin{Bmatrix} 2 & 2 & 0 \\ I_i & I_i & I_j \end{Bmatrix}, \quad (10)$$

where $\mathcal{M}(E2)$ is the transition matrix element and {} is a $6j$ symbol. While the sum extends over the complete set of states I_j, it generally is determined by a few key matrix elements. The Q^2 invariant extracted from the experimental data, $\langle Q^2 \rangle$, can be related to the β shape parameter

$$\langle Q^2 \rangle = q_0^2 \langle \beta^2 \rangle, \quad (11)$$

with $q_0 = \frac{3}{4\pi} Z R_0^2$ with $R_0 = 1.2A^{\frac{1}{3}}$ fm. Focusing on the ground state and the 0^+ intruder band head, $\langle Q^2 \rangle$ parameters extracted [19] from the data are 0.53(1) and 1.1(1)

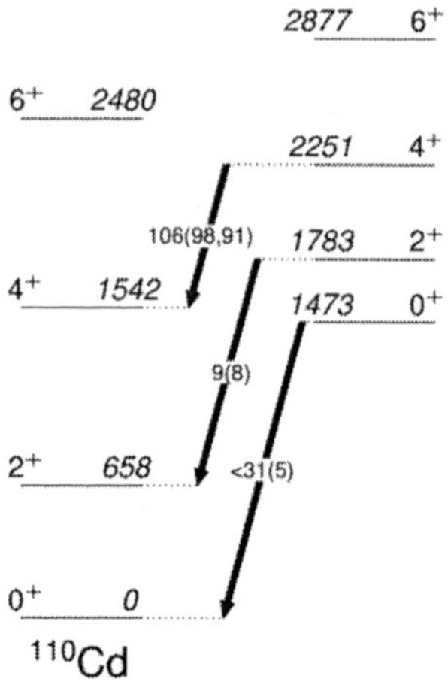

Figure 8. Partial level scheme for ^{110}Cd showing the $\rho^2(E0)\cdot 10^3$ values for transitions from the deformed intruder band to the ground-state band. The upper limit for the $0^+ \to 0^+$ transition is due to a lower limit on the lifetime for the 1473-keV 0^+ level.

e^{2}b^2, respectively, leading to $\langle\beta\rangle$ values of 0.187(2) and 0.27(1). In ^{114}Cd, $\rho^2(E0; 0_2^+ \to 0_1^+) = (19 \pm 2)\times 10^{-3}$, and with the above values of β, Eqn.6 leads to an admixture of $a^2 = 0.08$ of the intruder 0^+ wave function in the ground state.

7 Conclusions

Conversion electron spectroscopy is a powerful tool for the elucidation of level schemes, especially those involving highly-converted, low-energy transitions that are not always observed with γ-ray detectors. Furthermore, they provide a way to measure the $E0$ components of $J^\pi \to J^\pi$ transitions. The PACES array, employed at TRIUMF-ISAC with the 8π and GRIFFIN spectrometers, has provided such data. Examples of electron data from the decay of ^{124}Cs^m have been given that led to a modification of the decay scheme of this isomer. Data from the decay of ^{110}In were used to extract new $E0$ values, and a new $\rho^2(E0)$ value was determined for the decay of the 4^+ deformed intruder band to the 4^+ member of the ground-state band. Combining data from a previous Coulomb excitation study of ^{114}Cd with the available $\rho^2(E0)$ values permitted the admixture of the intruder 0^+ state in the ground state to be determined.

Acknowledgements

This work was supported in part by the Natural Sciences and Engineering Research Council (Canada) and the U.S. National Science Foundation under Grant No. PHY-1305801. TRIUMF receives federal funding via a contribution agreement through the National Research Council of Canada.

References

[1] P. E. Garrett *et al.*, J. Phys. Conf. Ser. **639** 012006 (2015).
[2] P. E. Garrett *et al.*, Nucl. Instrum. Meth. B **261** 1084 (2007).
[3] A. B. Garnsworthy and P. E. Garrett, Hyper. Int. **225** 121 (2014).
[4] C. E. Svensson and A. B. Garnsworthy, Hyper. Int. **225** 127 (2014).
[5] P. E. Garrett, Hyper. Int. **225** 137 (2014).
[6] P. E. Garrett *et al.*, Phys. Rev. C **86** 044304 (2012).
[7] A. Diaz Varela, P. E. Garrett, G. C. Ball *et al.*, E. Phys. J. Conf. **66** 02029 (2014).
[8] B. Jigmeddorj, P. E. Garrett, A. Diaz Varela *et al.*, JPS Conf. Proc. **6** 030014 (2015).
[9] B. Jigmeddorj, P. E. Garrett, A. Diaz Varela *et al.*, Eur. Phys. J. **A52**, 36 (2016).
[10] K. L. Green, P. E. Garrett, R. A. E. Austin *et al.*, Phys. Rev. C **80** 032502 (2009).
[11] P. E. Garrett, K. L. Green, R. A. E. Austin *et al.*, Acta Phys. Pol. **B42** 799 (2011).
[12] B. Jigmeddorj, P. E. Garrett, C. Andreoiu *et al.*, E. Phys. J. Conf. **107** 03014 (2016).
[13] A. J. Radich *et al.* Phys. Rev. C **91** 044320 (2015).
[14] J. L. Wood, E. F. Zganjar, C. De Coster, and K. Heyde, Nucl. Phys. **A651** 323 (1999).
[15] T. Kibédi, and R. H. Spear, Atom. Data Nucl. Data Tables **89** 77 (2005).
[16] J. P. Davidson, Rev. Mod. Phys. **37**, 105 (1965).
[17] K. Kumar, Phys. Rev. Lett. **28**, 249 (1972).
[18] D. Cline, Ann. Rev. Nucl. Part. Sci. **36**, 683 (1986).
[19] C. Fahlander, A. Bäcklin, L. Hasselgren*et al.*, Nucl. Phys. **A485** 318 (1988).

EPJ Web of Conferences **123**, 02006 (2016)
Heavy Ion Accelerator Symposium 2015 DOI: 10.1051/epjconf/201612302006

Investigating shape evolution and the emergence of collectivity through the synergy of Coulomb excitation and β decay

J.M. Allmond[1]

[1] *Physics Division, Oak Ridge National Laboratory, Oak Ridge, Tennessee 37831, USA*

Abstract. The synthesis of Coulomb excitation and β decay offers very practical advantages in the study of nuclear shapes and collectivity. For instance, Coulomb excitation is unique in its ability to measure the electric quadrupole moments, i.e., $\langle I_2^\pi || M(E2) || I_1^\pi \rangle$ matrix elements, of excited, non-isomeric states in atomic nuclei, providing information on the intrinsic shape. However, the Coulomb excitation analysis and structural interpretation can be strongly dependent upon weak transitions or decay branches, which are often obscured by the Compton background. Transitions of particular interest are those low in energy and weak in intensity due to the E_γ^5 attenuation factor. These weak decay branches can often be determined with high precision from β-decay studies. Recently, ^{106}Mo and ^{110}Cd were studied by both Coulomb excitation and β decay. Preliminary results of new weak decay branches following β decay of ^{110m}Ag to ^{110}Cd are presented; these results will challenge competing interpretations based on vibrations and configuration mixing.

1 Introduction

Low-energy decay branches of excited states are often weaker than alternative decay paths with higher energy; this is due to an energy-dependent attenuation factor. Even decays with large reduced transition probabilities, e.g., $B(E2)$, can appear weak when attenuated, $I_\gamma \propto B(E2) \times E_\gamma^5$. The $B(E2)$ magnitudes of these weak, low-energy transitions can often provide a way to differentiate between competing interpretations and calculations. However, these weak transitions can be difficult to observe. γ-γ coincidence measurements following β decay can often be used to measure weak decay branches with high precision, cf. [1]. The absolute $B(E2)$ magnitudes or $\langle I_2^\pi || M(E2) || I_1^\pi \rangle$ matrix elements can then be determined from Coulomb excitation data by normalizing through the strong decay branches.[1]

^{106}Mo and ^{110}Cd were recently studied by both Coulomb excitation and β decay. New weak decay branches of ^{110}Cd following β decay of ^{110m}Ag are presented. The ^{110}Cd study is largely motivated by a previous Coulomb-excitation study of ^{114}Cd [2] and recent evidence against a vibrational interpretation of nuclei in the Cd region based on measurements of lifetimes and intensities of weak decay branches [3–8].

[1]This manuscript has been authored by UT-Battelle, LLC, under Contract No. DE-AC05-00OR22725 with the U.S. Department of Energy. The United States Government retains and the publisher, by accepting the article for publication, acknowledges that the United States Government retains a non-exclusive, paid-up irrevocable, world-wide license to publish or reproduce the published form of the manuscript, or allow others to do so, for United States Government purposes.

2 Experiment and Results

A ^{110m}Ag(249.83 d) source was made at the High Flux Isotope Reactor (HFIR) at Oak Ridge National Laboratory (ORNL) by activating a natural Ag foil. The foil was then placed into the center of the γ-ray detector array CLARION [9], consisting of 11 Compton-suppressed HPGe Clover detectors at a distance of 21.75 cm with a total photo-peak efficiency of 3% at 1 MeV. This investigation focused on high-precision, ultra-high statistics measurements of weak γ-ray transitions between excited states.

Nearly 3 billion, 3×10^9, γ-γ coincidence events were recorded. A sample γ-ray spectrum is shown in Fig. 1, which reveals 9 new transitions in coincidence with the 1783-keV transition (i.e., a 1.1×10^{-4} decay branch [10]). These new transitions are very weak. For instance, the 379-keV transition has a decay branch of roughly 1 in 200,000 β decays.

The branching ratios and transition strengths of the γ rays from these data will test the so called "strong vibration-intruder mixing" scenario [11–13] used to explain the anomalous 0_2^+ and 0_3^+ decay strengths. Within this interpretation, the transitions outlined in Fig. 1 are an extensive set of new transitions from weakly collective states to intruder states. Based on the selective population of the (^{3}He,n) reaction [14], these intruder-state candidates are thought to be of π 2p-4h nature [15]. Most importantly, the new transitions, when combined with the Coulomb excitation data, will further complete the model-independent Kumar-Cline sum rules (shape invariants) [16] of the low-lying, non-yrast states.

DOI: 10.1051/epjconf/201612302006

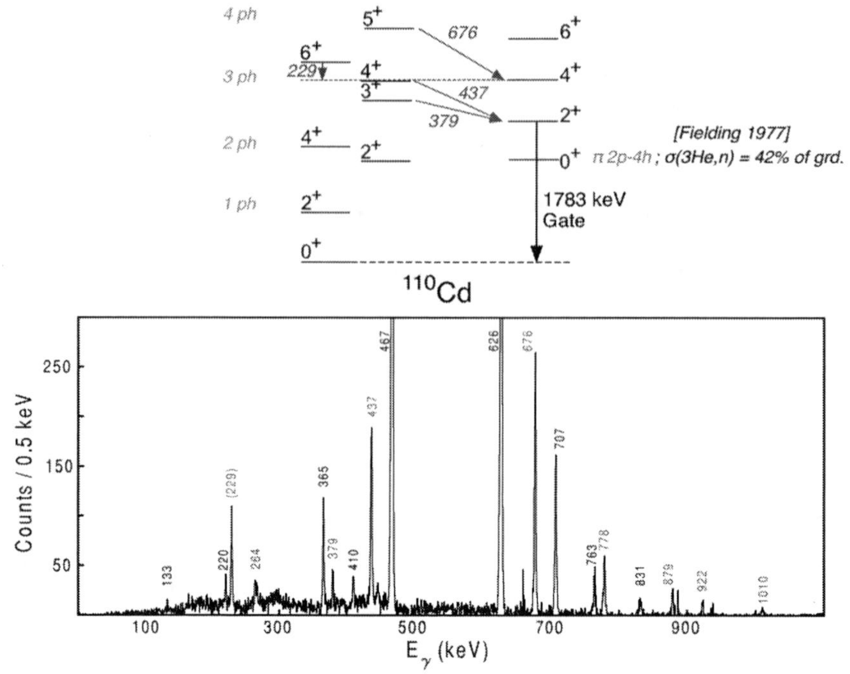

Figure 1. A partial level scheme and γ-ray spectrum of ^{110}Cd following ^{110m}Ag β decay and a gate on the 1783-keV γ-ray transition (a 1.1×10^{-4} per β decay branch [10]). The new transitions are labeled by energy in red ; the 229-keV γ ray was recently reported for the first time by Garrett *et al.*, [6]. These new transitions are incredibly weak. For instance, the 379-keV transition has a decay branch of roughly 1 in 200,000 β decays.

3 Conclusions

When these decay data are coupled with multiple-step Coulomb excitation data, the result will be an unprecedented set of electromagnetic moments for a weakly collective nucleus; this is an essential step in elucidating the nature of emergent collectivity.

Acknowledgements

The author gratefully acknowledges David Glasgow and the HFIR-ORNL operations staff for making the ^{110m}Ag source. This material is based upon work supported by the U.S. Department of Energy, Office of Science, Office of Nuclear Physics and this research used resources of the High Flux Isotope Reactor of Oak Ridge National Laboratory, which is a DOE Office of Science User Facility.

References

[1] W. D. Kulp *et al.*, Phys. Rev. C **73**, 014308 (2006).
[2] C. Fahlander *et al.*, Nucl. Phys. **A485**, 327 (1988).
[3] P. E. Garrett *et al.*, Phys. Rev. C **75**, 054310 (2007).
[4] P. E. Garrett *et al.*, Phys. Rev. C **78**, 044307 (2008).
[5] P. E. Garrett and J. L. Wood, J. Phys. G: Nucl. Part. Phys. **37**, 064028 (2010).
[6] P. E. Garrett *et al.*, Phys. Rev. C **86**, 044304 (2012).
[7] J. C. Batchelder *et al.*, Phys. Rev. C **80**, 054318 (2009).
[8] J. C. Batchelder *et al.*, Phys. Rev. C **86**, 064311 (2012).
[9] C. J. Gross *et al.*, Nucl. Instrum. Methods Phys. Res. A **450**, 12 (2000).
[10] Evaluated Nuclear Structure Data File (ENSDF), http://www.nndc.bnl.gov/ensdf/.
[11] K. Heyde, P. Van Isacker, M. Waroquier, G. Wenes, and M. Sambataro, Phys. Rev. C **25**, 3160 (1982).
[12] J. Jolie and K. Heyde, Phys. Rev. C **42**, 2034 (1990).
[13] M. Deleze *et al.*, Nucl. Phys. **A551**, 269 (1993).
[14] H. W. Fielding *et al.*, Nucl. Phys. **A281**, 389 (1977).
[15] K. Heyde and J. L. Wood, Rev. Mod. Phys. **83**, 1467 (2011).
[16] K. Kumar, Phys. Rev. Lett. **28**, 249 (1972); D. Cline, Annu. Rev. Nucl. Part. Sci. **36**, 681 (1986).

EPJ Web of Conferences **123**, 02007 (2016)
Heavy Ion Accelerator Symposium 2015

DOI: 10.1051/epjconf/201612302007

Decay spectroscopy of exotic nuclei at RIBF

H. Watanabe[1,2]

[1] *IRCNPC, School of Physics and Nuclear Energy Engineering, Beihang University, China*
[2] *RIKEN Nishina Center*

Abstract. Neutron-rich nuclei around the doubly magic nucleus ^{132}Sn and the double midshell nucleus ^{170}Dy have been investigated by means of decay spectroscopy techniques with the EURICA setup at the RIBF facility at RIKEN. The nuclei of interest were produced by in-flight fission of a high-intensity ^{238}U beam at 345 MeV/u. In this contribution, some selected topics are reported.

1 Introduction

Exotic nuclei far from the valley of β stability can serve as a good probe for studying new phenomena, such as neutron skin/halo structures and appearance/disappearance of magic numbers, which emerge with largely asymmetric values of the N/Z ratio. The advent of the third-generation in-flight fragment separator facility, the RI-Beam Factory (RIBF) at RIKEN [1], enables systematic investigations of such novel nuclear properties and allows for extension of our knowledge of nuclear structures towards the neutron and proton drip lines. In that occasion, γ-ray spectroscopy following the β decay is an effective tool for probing low-lying yrast and non-yrast states, from which key information on nuclear structure, such as shape transitions/coexistence and single-particle orbits, can be obtained. For the study of rare isotopes, especially when the nucleus of interest lies at the boundaries of availability for spectroscopic studies, isomeric decays are likely to be a more useful means than β decays to populate excited levels. The combined β-γ and isomeric-decay measurements using state-of-the-art detection systems at RIBF, which has the capability of providing the world's strongest RI beams, are at the forefront of exploration of exotic nuclei far from stability.

The RIBF was commissioned in 2007, with the adoption of a sequential acceleration system consisting of a linac injector (RILAC) and four ring cyclotrons (RRC-fRC-IRC-SRC), which can provide heavy-ion beams at 345 MeV/u. The primary-beam intensity increases year by year owing to the steady improvement of the accelerators and the related devices. For instance, the intensity of a ^{238}U beam was lower than 0.1 pnA when the facility was commissioned, but now it can be delivered with a stable intensity higher than 20 pnA. The first decay spectroscopy experiment at RIBF was carried out in 2009 with four Compton-suppressed Clover-type Ge detectors arranged around an active stopper system consisting of nine double-sided silicon-strip detectors (DSSSD) stacked compactly. A major aim of this work was the study of neutron-rich nuclei around $Z = 40$ and $A = 110$, where the shape transitions from prolate, via γ-soft, to oblate deformations are predicted to occur with increasing the number of neutrons. New results obtained include the observation of shape evolution in ^{106,108}Zr [2], a possible oblate-shape isomer in ^{109}Nb [3], and a large-amplitude γ-soft dynamics in ^{110}Mo [4]. These findings revealed the shape-transitional phenomena in this neutron-rich region for the first time.

Research opportunities for decay spectroscopy at RIBF can be expanded in the EURICA (EUROBALL-RIKEN Cluster Array) project [5]. The main body of EURICA consists of 12 Cluster-type HPGe detectors, surrounding a highly segmented silicon stopper system named WAS3ABi [5]. In addition to the normal EURICA setup, 18 $LaBr_3$(Ce) detectors and plastic scintillators are installed for the fast-timing measurement of γ and β rays, respectively. The aim of EURICA is to pin down currently controversial subjects in nuclear physics and nuclear astrophysics, such as the evolution of shell structures that can lead to the appearance or disappearance of the spherical magic numbers, the effect of weak binding and pairing in largely diffused neutron densities, search for stable oblate, triaxial, and higher-order deformations at low excitation energy, and the decay properties of neutron-rich isotopes relevant to the r-process nucleosynthesis. A wide range of unstable nuclei on the Segre chart are within the scope of EURICA. In this report, some selected results of the EURICA experimental campaigns will be introduced, highlighting γ-ray spectroscopy of exotic nuclei in the vicinity of doubly magic ^{132}Sn and doubly mid-shell ^{170}Dy.

2 Selected topics from EURICA

2.1 Isomeric states in $^{126}Pd_{80}$ and $^{128}Pd_{82}$

Figure 1(a) shows a γ-ray energy spectrum measured in delayed coincidence with ^{128}Pd ions. Four γ rays at ener-

DOI: 10.1051/epjconf/201612302007

gies of 75, 260, 504, and 1311 keV have been unambiguously observed. These γ rays are found to be in mutual coincidence and exhibit consistent time behavior. Therefore, we conclude that they proceed through a single cascade originating from one isomeric state. A least-squares fit of the summed gated time spectra of the isomeric-decay transitions yields $T_{1/2} = 5.8(8)$ μs. The total internal conversion coefficient (α_T) for the 75-keV transition derived from a comparison with the 1311-keV γ-ray intensity is 2.6(17), which is consistent with the theoretical value of $\alpha_T = 3.88$ for an $E2$ multipolarity. Based on these arguments, the level scheme of ^{128}Pd is proposed as displayed in Fig. 2, where the spin and parity of the 5.8-μs isomeric state at 2151 keV is assigned as $J^\pi = 8^+$. A transition strength of $B(E2; 8^+ \to 6^+) = 0.22(3)$ W.u. can be obtained from the measured half-life of the 2151-keV isomeric state.

Assuming that the excited states in ^{128}Pd are the same multiplet members (with seniority $\upsilon = 2$, as will be discussed later), the half-life of the $J^\pi = 6^+$ state is estimated to be 22(3) ns from that of the 8^+ isomer. Since this value is significantly shorter than the flight time of fragments from the production target to the implantation position (~ 650 ns), there would be little ^{128}Pd ions remaining at this state during the flight. Hence, the observed intensities of the four cascade γ rays should depend only on the population of the 8^+ isomeric state.

The excitation energies of the $J^\pi = 2^+ - 8^+$ states in ^{128}Pd are comparable to those in ^{130}Cd [6]. The constancy of level energies is characteristic of the seniority scheme, where seniority υ counts the number of nucleons that are not in pairs coupled to spin zero. In case of a n-particle (or n-hole) system in a single-j shell, the level energies with identical J^π and υ are independent of n. Thus, the excited states in ^{128}Pd can be interpreted in terms of the $\upsilon = 2$ configuration of the $\pi g_{9/2}$ subshell, which locates just below the $Z = 50$ shell gap.

A direct implication of the $N = 82$ shell closure can be corroborated from the behavior of the $E2$ transition strength for the $8^+ \to 6^+$ isomer decay in ^{128}Pd. The $E2$ matrix elements between $\upsilon = 2$ states are close to zero near the middle of the valence shell in semi-magic nuclei, giving rise to seniority isomerism. For the $N = 82$ isotones, it turns out that the $B(E2; 8^+ \to 6^+)$ value is much smaller in ^{128}Pd than in ^{130}Cd, as expected in the exact seniority classification, namely, $B(E2; \upsilon = 2, n = 4) = \frac{1}{9}B(E2; \upsilon = 2, n = 2)$. This observation indicates that both the $J^\pi = 8^+$ and 6^+ states in the $N = 82$ isotones have good seniority $\upsilon = 2$ in the well isolated $\pi g_{9/2}$ subshell, in consequence of the robust shell closure.

The level scheme of ^{126}Pd constructed in the current work is also shown in Fig. 2. The 693-keV γ ray, which is the most intense peak of those observed in delayed coincidence with ^{126}Pd ions [Fig. 1(b)], is assigned as the $2^+ \to 0^+$ transition. The coincidence spectrum with a gate on the 542-keV γ ray clearly exhibited γ rays at energies of 693, 788, and 86 keV, but not at 1330 keV, indicating that two parallel cascades stem from a common level at 2023 keV. The 86-keV transition is placed just above the 2023-keV level, because this γ ray was observed to precede all the four transitions.

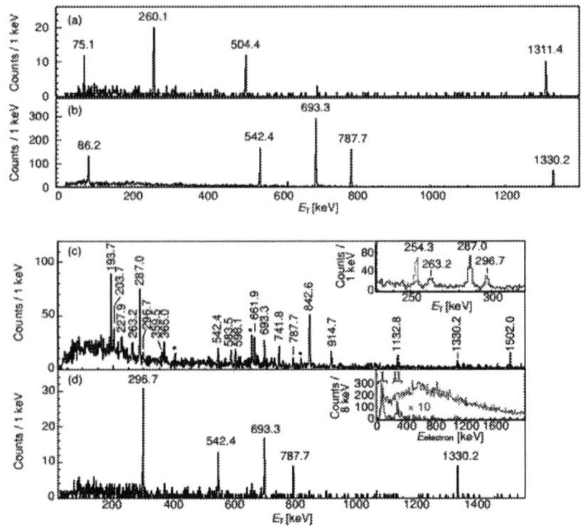

Figure 1. Gamma-ray spectra measured (a) in coincidence with ^{128}Pd ions within a time range of 0.15–25 μs, (b) in coincidence with ^{126}Pd ions within 0.15–5 μs, and (c) within 50 ms after the ^{126}Pd implantation with a gate on an electron-γ time difference of $-0.5 \le \Delta t_{e\gamma} \le 0.5$ μs. Contaminants from the granddaughter ^{126}Cd are marked with filled circles. The inset magnifies the energy region from 230 to 310 keV measured with $-4.0 \le \Delta t_{e\gamma} \le 0.5$ μs (red) and $-0.5 \le \Delta t_{e\gamma} \le 50$ μs (blue). (d): γ rays measured with $-4.0 \le \Delta t_{e\gamma} \le 0.5$ μs, and additionally, with a gate on the electron peak marked with "I" in the inset, where the spectrum depicted with the red line (multiplied by a factor of ten) is obtained with a sum of gates on the γ rays below the (5^-) isomer in ^{126}Pd, while the blue line represents an electron spectrum without γ-ray gates.

A half-life of 0.44(3) μs is derived from the time distribution of the 86-keV γ ray relative to the beam implantation, while the time difference between the 86-keV γ ray and the subsequent transitions provides $T_{1/2} = 0.33(4)$ μs. Thus, it is found that two isomeric states lie at 2023 and 2110 keV. The assignment of spins and parities for the two isomeric states is based on the γ-ray feeding pattern and transition probabilities. Given the $J^\pi = 5^-$ and 7^- assignments for the 2023- and 2110-keV levels, respectively, the strengths of the isomeric decay transitions are $B(E1; 5^- \to 4^+) = 2.9(4) \times 10^{-9}$ W.u. (542 keV), $B(E3; 5^- \to 2^+) = 0.24(3)$ W.u. (1330 keV), and $B(E2; 7^- \to 5^-) = 2.13(14)$ W.u. (86 keV). These strengths are comparable with the values of the corresponding transitions in the $N = 80$ isotone ^{128}Cd [7].

The γ rays below these isomers, except for the 86-keV line, have been also observed in coincidence with electrons that were associated with the prior implantation of ^{126}Pd, as demonstrated in Fig. 1(c). This observation implies the existence of a long-lived, higher-spin isomer which decays via the cascades that include electromagnetic transitions with relatively large total conversion coefficients. With gates on these γ rays, a prominent peak can be found in an electron spectrum [marked with "I" in the inset of

DOI: 10.1051/epjconf/201612302007

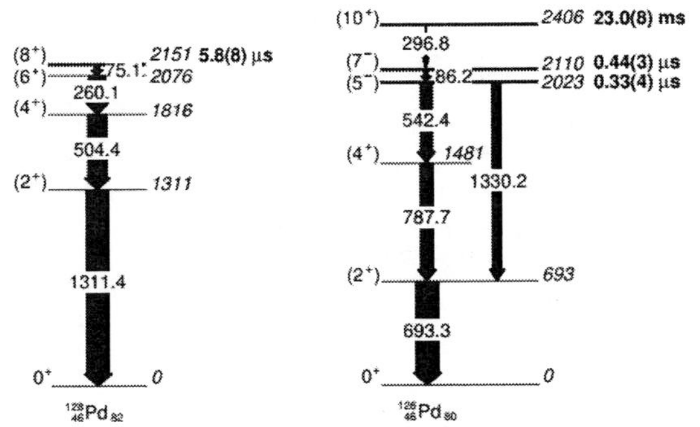

Figure 2. Partial level schemes of ^{128}Pd (left) and ^{126}Pd (right) established in the present work.

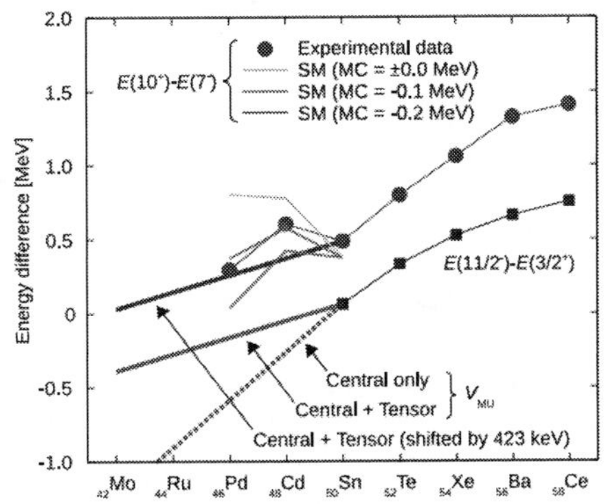

Figure 3. (color online). Energy differences between the 10^+ and 7^- states, $\Delta E_{7^-}^{10^+}$, in $N = 80$ (filled circles) and between the $11/2^-$ and $3/2^+$ states in $N = 81$ (filled squares). For $Z \le 50$, the differences in energy between the $1h_{11/2}$ and $2d_{3/2}$ neutron orbits calculated by V_{MU} interaction [8] and $\Delta E_{7^-}^{10^+}$ by shell-model (SM) calculations with monopole corrections (MC) are also depicted.

Fig. 1(d)]; this corresponds to the conversion electrons for the 86-keV, $E2$ transition ($\alpha_T = 2.374$). In Fig. 1(d), a γ-ray at 297 keV is clearly visible in addition to the γ rays below the (5^-) isomer by gating on the 86-keV IC peak. The appearance of the 297-keV peak is emphasized by taking a γ-ray time condition earlier than electron events, as is evident from the inset of Fig. 1(c), suggesting that this new γ ray precedes the highly converted 86-keV transition. Thus, the long-lived isomer can be identified at an excitation energy of 2406 keV. A peak marked with "II" in the inset of Fig. 1(d) is expected to arise from the conversion electrons for the 297-keV transition, being most likely of an $E3$ character ($\alpha_T = 0.1197$). In addition to the internal-decay branch, β decay from the long-lived isomer was observed to populate excited states at high spins in ^{126}Ag. The isomeric half-life was determined to be 23.0(8) ms by taking a weighted average of those derived from the time distributions of the internal- and external-decay branches.

The main surprise in the present work is the small energy difference between the (10^+) and (7^-) isomers, $\Delta E_{7^-}^{10^+}$, in ^{126}Pd, compared to the analogous levels in the heavier $N = 80$ isotones, as demonstrated in Fig. 3. Since these two levels consist predominantly of maximally aligned two neutron-hole configurations, $(\nu 1h_{11/2}^{-2})_{10^+}$ and $(\nu 1h_{11/2}^{-1} 2d_{3/2}^{-1})_{7^-}$, their level energies depend on the single-particle energies (SPEs) of the $\nu 1h_{11/2}$ and $\nu 2d_{3/2}$ neutron orbits, as well as the strength of interactions between them. It can be seen in Fig. 3 that the energy of the $11/2^-$ level relative to the $3/2^+$ ground state in the neighboring $N = 81$ isotones decreases when approaching $Z = 50$. Note that these states are of one neutron-hole nature, and that the smooth reduction in energy can be essentially interpreted in terms of a short-range proton-neutron interaction as follows: The $\pi 1g_{7/2}$ proton orbit lies just above the $Z = 50$ shell closure. The monopole interaction between $\pi 1g_{7/2}$ and $\nu 1h_{11/2}$, a spin-flip pair with $\Delta\ell = 1$ and $\Delta n = 0$, is stronger than the $\pi 1g_{7/2}$-$\nu 2d_{3/2}$ pair, due to the larger overlap between the radial wave functions of the two orbits. Therefore, when emptying the $\pi 1g_{7/2}$ subshell, the $\nu 1h_{11/2}$ orbit is relatively less bound than the $\nu 2d_{3/2}$ one. The moderation in the slope at higher Z may be ascribed to the partial occupation of the $\pi 2d_{5/2}$ orbit, which exerts a weaker (stronger) monopole interaction on the $\nu 1h_{11/2}$ ($\nu 2d_{3/2}$) neutrons than $\pi 1g_{7/2}$. As is evident from the comparison in Fig. 3, a similar trend is observed for $\Delta E_{7^-}^{10^+}$. This finding supports that these isomers can serve as sensitive probes for the evolution of the constituent neutron shell orbits.

Below $Z = 50$, protons in the $\pi 1g_{9/2}$ subshell play a major role in changing the neutron SPEs, because this orbit is near the Fermi surface. The neutron SPEs are estimated by the monopole-based universal interaction, V_{MU}, which consists of the Gaussian central force and the tensor force based on the π+ρ meson exchanges, using the parameters fixed in Ref. [8]. The V_{MU} interaction has been applied to other regions of exotic nuclei and successfully reproduced various types of the shell evolution [8]. Figure 3 shows the evolutions of the relative $11/2^-$ energy starting from the experimental value in ^{131}Sn, predicted by the V_{MU} calculation with $A = 130$. In the calculation, the $\pi 1g_{9/2}$ subshell is considered to be well isolated between $Z = 50$ and 40. In Fig. 3, only with the central force the $11/2^-$ energy rapidly decreases to negative values as the proton number decreases, indicating that the $\nu 1h_{11/2}$ orbit moves away from the $\nu 2d_{3/2}$ one towards the $N = 82$ gap. If there is the tensor force effect, however, the slope becomes less steep. The observed reduction in $\Delta E_{7^-}^{10^+}$ from ^{130}Sn to ^{126}Pd is consistent with the expectation of V_{MU} including both the central and tensor forces, given the same trend as the $11/2^-$ relative energies. Based on these findings, it can be concluded that the tensor force slows down the upward drift of the $\nu 1h_{11/2}$ SPE when protons are removed from the $\pi 1g_{9/2}$ orbit below ^{132}Sn.

 DOI: 10.1051/epjconf/201612302007

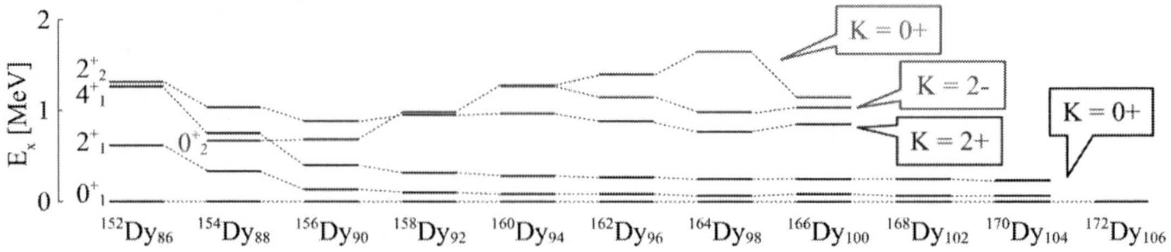

Figure 4. Systematic of the low-lying levels in even-even Dy isotopes.

2.2 Decay spectroscopy of neutron-rich rare-earth isotopes

It is well known that rare-earth isotopes beyond $N = 90$ have well-deformed prolate shapes. According to the fundamental phenomenon of the nuclear collectivity being enhanced as the number of valence nucleons increases, it is conjectured that the maximum ground-state deformation occurs in dysprosium-170 (^{170}Dy: $Z = 66$, $N = 104$), which lies at the middle of the major shells for both proton and neutron between the well-established doubly closed-shell isotopes ^{132}Sn ($Z = 50$, $N = 82$) and ^{208}Pb ($Z = 82$, $N = 126$). In the actual nuclei, however, the stability of shape should be influenced by the presence of sub-shell closures, residual interactions, and possibly, the neutron excess. Therefore, well-deformed nuclei in this neutron-rich region can serve as a benchmark for testing nuclear mean-field model calculations.

Concerning astrophysical interest, it has been suggested that the decay properties of deformed nuclei in this neutron-rich region, such as neutron emission probabilities and β-decay rates, play a significant role in forming a pygmy peak around $A = 160$ in the r-process solar abundance distribution [9]. This is the so-called the rare-earth element (REE) peak, which appears between the prominent r-process peaks at $A \approx 130$ and 195. While it is understood that the main r-process peaks are due predominantly to the "waiting point" nuclei arising from the competition between slow neutron-capture rates and long β-decay lifetimes at the neutron closed shells, the REE peak appears far away both from the proton and neutron shell closures, suggesting the presence of a different mechanism for its formation. In this context, the nuclear deformation in this neutron-rich region is considered to significantly affect the REE peak formation. However, it is still unclear how the nuclear shape evolves and where the deformation maximum takes place in the double midshell region because spectroscopic information on the excited states remains scarce due to the difficulties in access to this neutron-rich region. Theoretical models that are often used as input for r-process calculations can not reproduce the experimentally measured deformations very well. Thus, the REE peak can serve as a unique probe of the r-process nucleosynthesis, despite its less abundance than the other major peaks at $A \approx 130$ and 195, and better knowledge on the evolution of collectivity in this region is required.

The major aim of the present work is to scrutinize the structural evolution of Dy isotopes in the double midshell region by means of γ-ray spectroscopy following isomer and β decays, with particular attention to the physics cases and observables characteristic of deformed nuclei as outlined below:

Low-lying levels in even-even isotopes

Spectroscopic study of low-lying levels in even-even nuclei provides significant information on nuclear shapes and the resulting collective motions. According to the global systematics, it is empirically known that the deformation parameter β is inversely proportional to the square root of the first 2^+ level energy $E(2^+_1)$ [10], while the $R_{4/2} = E(4^+_1)/E(2^+_1)$ ratio represents the rigidity limits of quadrupole collectivity being either rotational (~ 3.33) or vibrational (~ 2.0). Figure 4 shows the evolution of the low-spin states in the Dy isotopic chain. Inspection of the 2^+_1 and 4^+_1 states indicates that the vibrational $\rightarrow$ rotational phase transition takes place around $A = 88$, and the $R_{4/2}$ values approach the rigid-rotor limit towards the middle of the neutron shell.

Low-lying non-yrast states are sensitive probes for microscopic structure of deformed nuclei. In general, the second 2^+ states (2^+_2) are associated predominantly with the bandhead of the γ-vibrational band. A qualitative explanation for the occurrence of the γ-vibrational $K^\pi = 2^+$ state at relatively low excitation energy can be given in terms of the single-particle orbits in the deformed potentials; the operator r_2Y_{22} acts on the single-particle states occupying a specific Nilsson orbit $\Omega[Nn_z\Lambda]$ which satisfies the selection rule, $\Delta N = \Delta n_z = 0$ and $\Delta\Lambda = \Delta\Omega = \pm 2$, giving rise to the γ-vibrational motions in the collective dynamics. In the region of interest, there are several pairs of orbits that meet the r_2Y_{22} selection rule near the proton and neutron Fermi surfaces. The wave function of the 2^+_2 state consists of a number of such two-quasiparticle excitations, which coherently contribute to the γ-vibrational amplitudes. Meanwhile, the interpretation of the excited 0^+ states (0^+_2) is rather complicated because there are several origins of the $K^\pi = 0^+$ excitations like the β vibration, pairing excitations, two-γ-phonon vibration, and intruder states. The $K^\pi = 2^-$ states observed in the even Dy isotopes are interpreted as the band head of the octupole vibrational band with the predominant component $\pi^2 7/2^-[523] \otimes 3/2^+[411]$.

K-forbidden decays from multi-quasiparticle states (K isomers)

The region of well-deformed nuclei around $A = 170-180$ ($Z \sim 70$, $N \sim 110$) is known for the existence of lots of K isomers [11, 12], including those with half-lives longer than one hour. K quantum number is defined as the projection of the total nuclear spin on the symmetry axis of the deformed nucleus. In addition to the spin selection rule, electromagnetic transitions connecting the states with different K values (ΔK) are required to fulfill a selection rule that the multipole order of the transition λ is equal to or greater than ΔK, namely, $\lambda \geq \Delta K$. However, this K-conservation law is often violated due to various sources of symmetry breaking, such as Coriolis effects and softness in terms of the γ (axially-asymmetric) degree of freedom, and hence strongly hindered transitions can take place. Such K-forbidden transitions are evaluated practically in terms of a reduced hindrance $f_\nu = F^{1/\nu}$, where $F = T_{1/2}(\gamma)/T^W_{1/2}$ represents a hindrance relative to the Weisskopf single-particle estimate of the decay half-life, with the degree of K-forbiddenness defined by $\nu = \Delta K - \lambda$. It is known empirically that the value of f_ν ranges from 30 to 200 when K is well sustained for both the initial and final states.

High-Ω orbits near the Fermi surface can contribute to the emergence of high-K isomers at relatively low excitation energy. In general, a large (and robust) axial deformation yields good conditions for the formation of high-K isomeric states. Thus, the discovery and characterization of such high-K multi-quasiparticle states will shed light on various aspects of nuclear structure, such as intrinsic orbits near the Fermi surface, paring and other residual interactions, and the degree of axial symmetry in deformed nuclei. Furthermore, it is likely that high-K isomers decays through low-lying excited levels not only in the ground-state band, but also in deformed-vibrational bands which represent crucial information for understanding the evolution of deformed structure.

This work focuses particularly on isomeric states that are expected to appear in ^{170}Dy ($N = 104$) with $K^\pi = 6^+$ and in ^{172}Dy ($N = 106$) with $K^\pi = 8^-$. The near-stability $N = 104$ isotones are well known to have $K^\pi = 6^+$ isomers [13], which are interpreted as being based on the two-quasineutron configuration $\nu^2 7/2^-[514] \otimes 5/2^-[512]$. Configuration-constrained potential-energy-surface calculations for the $K^\pi = 6^+$ configuration predict the analogous isomer at about 1.2 MeV in ^{170}Dy [14]. The large deformation and high degree of axial symmetry, predicted for both the two-quasineutron $K^\pi = 6^+$ state and ground-band members to which the isomer decays, suggest that this nucleus may represent an extremely good example of a "pure" K isomer [14]. Considering the $N = 106$ isotones, $K^\pi = 8^-$ isomers which decay via an $E1$ transition to the yrast 8^+ state have been identified over a broad range of proton number [13]. The $K^\pi = 8^-$ configuration is interpreted as the two-quasineutron configuration $\nu^2 7/2^-[514] \otimes 9/2^+[624]$. Based on the systematic behavior of the hindered $E1$ decays [15], it is expected that the $E1$ transition which de-excites the $K^\pi = 8^-$ isomer in

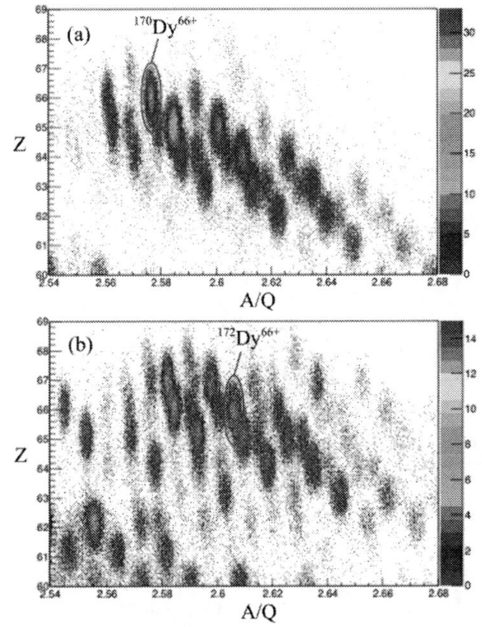

Figure 5. Particle identification spectra obtained with slit positions optimized for (a) ^{170}Dy^{66+} and (b) ^{172}Dy^{66+} ions.

^{172}Dy will have a reduced hindrance of $f_\nu \approx 100$, resulting in a half-life of a few seconds. This is expected to be comparable to that of the ground state in this neutron-rich region, presumably leading to competition from β decay in the isomer decay.

Neutron-rich rare-earth nuclei around $A = 170$ were produced by in-flight fission of a ^{238}U^{86+} beam at 345 MeV/u with the intensity of 12 pnA on average, impinging on a beryllium target with a thickness of 5 mm. The nuclei of interest were separated through the BigRIPS separator based on their physical and magnetic properties [16]. For purification of the secondary beams, wedge-shaped aluminum degraders with thicknesses of 4.5 and 1.0 mm were installed at the first (F1) and second (F5) dispersive focal planes, respectively. Identification of particles with the atomic number (Z) and the mass-to-charge ratio (A/Q) was achieved on the basis of the ΔE-TOF-$B\rho$ method, in which the energy loss (ΔE), time of flight (TOF), and magnetic rigidity ($B\rho$) were measured using the focal-plane detectors on the beam line. The secondary beams were transported with two different settings of the slits on the beam line; one is optimized for ^{170}Dy^{66+}, and the other for ^{172}Dy^{66+}. The obtained particle-identification spectra are shown in Fig. 5. About 8.3×10^3 ^{172}Dy ions were identified through the BigRIPS spectrometer.

The identified particles were implanted into a highly segmented active stopper system, named WAS3ABi [5], which consisted of two double-sided silicon-strip detectors (DSSSD) stacked compactly. Each DSSSD had a thickness of 1 mm with an active area segmented into sixty and forty strips (1-mm pitch) on each side in the horizontal and vertical dimensions, respectively. The DSSSDs also

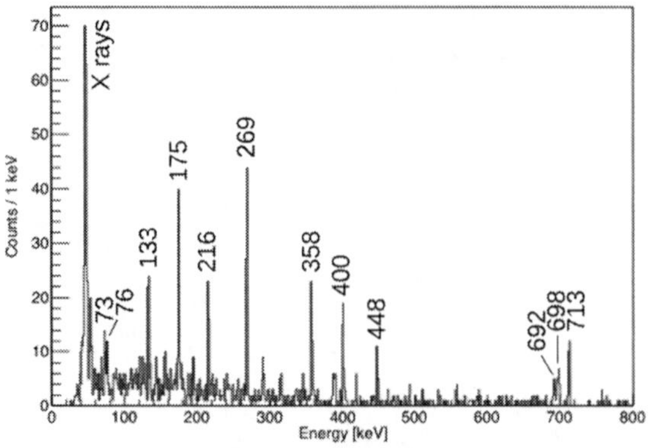

Figure 6. Gamma-ray spectrum measured in coincidence with electrons within 4 s after implantation of fully-stripped ^{172}Dy ions.

served as detectors for electrons following β-decay and internal conversion (IC) processes. Gamma rays were detected by the EURICA spectrometer [5], which consisted of twelve Cluster-type detectors, each of which contained seven HPGe crystals packed closely, which were developed for the former EUROBALL array. The γ-ray measurements were carried out with a time condition up to 100 μs relative to the trigger signal generated either from a plastic scintillation counter placed at the end of the beam line or from WAS3ABi.

The beam, electron, and γ-ray events were time-stamped and recorded by independent data-acquisition systems. For the analysis of beam-γ delayed coincidence, the γ-ray data sets were combined with those of the beam particles on an event-by-event basis using information on the time stamp. Isomeric states with (sub)microsecond lifetimes were identified with appropriate time gates. Meanwhile, all data sets containing beam, electron, and γ-ray events were used for β-γ and IC-γ coincidence analyses, in which the implantation of an identified particle was associated with the subsequent electron events that were detected in a given correlation area in the DSSSDs where the beam particle was implanted. Decay half-lives ($T_{1/2}$) in the millisecond range were extracted from the time distributions of γ-ray gated electron events with respect to the fragment implantation.

Figure 6 exhibits a γ-ray spectrum measured in coincidence with electrons within 4 seconds after implantation of fully-stripped ^{172}Dy ions. It is expected that the observed γ-ray spectrum consists of the transitions not only following the β decay from the ground state, but also deexciting the long-lived $K^{\pi} = 8^{-}$ isomer that is predicted for the $N = 106$ isotone as mentioned above. It will be possible to distinguish the internal and external decays by gating on time and energy spectra of electrons detected in the DSSSDs. The detailed analysis is still ongoing.

References

[1] Y. Yano, Nucl. Instrum. Methods B **261**, 1009 (2007).
[2] T. Sumikama, K. Yoshinaga, H. Watanabe *et al.*, Phys. Rev. Lett. **106**, 202501 (2011).
[3] H. Watanabe, T. Sumikama, S. Nishimura *et al.*, Phys. Lett. B **696**, 186 (2011).
[4] H. Watanabe, K. Yamaguchi, A. Odahara *et al.*, Phys. Lett. B **704**, 270 (2011).
[5] S. Nishimura, Prog. Theor. Exp. Phys. p. 03C006 (2012).
[6] A. Jungclaus, L. Cáceres, M. Górska *et al.*, Phys. Rev. Lett. **99**, 132501 (2007).
[7] L. Cáceres, M. Górska, A. Jungclaus *et al.*, Phys. Rev. C **79**, 011301 (2009).
[8] T. Otsuka, T. Suzuki, M. Honma *et al.*, Phys. Rev. Lett. **104**, 012501 (2010).
[9] R. Surman, J. Engel, J. R. Bennett *et al.*, Phys. Rev. Lett. **79**, 1809 (1997).
[10] S. Raman, C. Nestor and P. Tikkanen, At. Data Nucl. Data Tables **78**, 1 (2001).
[11] P. Walker and G. Dracoulis, Nature **399**, 35 (1999).
[12] F. G. Kondev, G. D. Dracoulis and T. Kibedi, At. Data Nucl. Data Tables **103-104**, 50 (2015).
[13] G. Dracoulis, G. Lane, F. Kondev *et al.*, Phys. Lett. B **635**, 200 (2006).
[14] P. H. Regan, F. R. Xu, P. M. Walker *et al.*, Phys. Rev. C **65**, 037302 (2002).
[15] G. D. Dracoulis, G. J. Lane, F. G. Kondev *et al.*, Phys. Rev. C **79**, 061303 (2009).
[16] T. Kubo, Nucl. Instrum. Methods B **204**, 97 (2003).

EPJ Web of Conferences **123**, 03001 (2016)
Heavy Ion Accelerator Symposium 2015 DOI: 10.1051/epjconf/201612303001

Nuclear structure effects on heavy-ion reactions with microscopic theory

K. Vo-Phuoc[1,a], C. Simenel[1], and E. C. Simpson[1]

[1] *Department of Nuclear Physics, Research School of Physics and Engineering, Australian National University, Canberra, ACT 2601, Australia*

Abstract. The self-consistent mean-field Hartree–Fock (HF) theory, both static and time-dependent (TDHF) versions, is used to study static and dynamic properties of fusion reactions between even $^{40-54}$Ca isotopes and ^{116}Sn. The bare nucleus-nucleus potential, calculated with the frozen HF approach, is affected by the ground-state density of the nuclei. However, once dynamical effects are included, as in TDHF, the static effects on the barrier are essentially washed out. Dynamic properties of the nuclei, including low-lying vibrational modes, are calculated with TDHF and selectively used in coupled-channels calculations to identify which modes have the most effect on the TDHF fusion threshold. Vibrations cannot fully explain the difference between the static HF and TDHF fusion barriers trend so other dynamical effects such as transfer are considered.

1 Introduction

Nuclear structure plays a major role in nuclear fusion reactions. Effects of internal nuclear structure on heavy-ion fusion can be seen by studying features of experimental fusion barrier distributions [1, 2]. Experiments have shown that dynamic effects such as low-lying vibrational couplings [3–5] and transfer reactions [6–10] strongly affect fusion reactions. Other dynamic effects, such as breakup [11–14] and rotational couplings [4], have also been seen to play a role.

Different theoretical approaches can be used to study fusion. Conventionally, the coupled-channels approach is applied to study reactions and compare fusion cross sections with experimental data, for example, but it requires input parameters of the structure of the nuclei in the reaction. When these parameters are not known or where experimental data is not available, a more microscopic theoretical method is valuable.

The Hartree–Fock (HF) method is a self-consistent theory which is a useful tool to study both nuclear structure and reactions as its only input is the nucleon-nucleon interaction. The time-dependent version (TDHF) allows for the study of dynamical effects on fusion reactions. TDHF has been used for several purposes for nuclear structure and reactions (see Refs. [15, 16] for reviews). Modern codes solve the TDHF equation in three dimensions and include spin-orbit interaction [17–20]. Recent applications include vibration [21–25], fusion [26–30], transfer [31–35], quasi-fission [36–38] and fission [39–42] studies.

TDHF includes many dynamical effects simultaneously and it is difficult to isolate the effect of a single dynamical coupling on a fusion reaction. Here the method of including microscopic inputs to coupled-channels (CC) analysis [27] is used to help unravel the TDHF dynamics and the influence of these on the fusion reactions.

2 Static effects

2.1 Numerical details

There are two steps to obtaining a nucleus-nucleus potential using the frozen HF method. The first step is the calculation of an individual HF ground state nucleus wavefunction. This was done for all the projectiles $^{40-54}$Ca and target ^{116}Sn. The HF method assumes that nucleons in a nucleus are independent particles and that they generate a self consistent mean-field. The ground state nucleus wavefunction for each nucleus was calculated using the ev8 code [43]. The HF equation

$$[h[\rho], \rho] = 0, \quad (1)$$

where ρ is the one-body density matrix of the independent particle system and $h[\rho]$ is the HF hamiltonian, is solved on a 3D grid on a box size $(2 \times 11.2)^3$ fm^3 with three planes of symmetry. The effective nucleon-nucleon interaction was taken to be the Skyrme interaction [44] with the SLy4d [17] parametrisation. This parametrisation ignores centre of mass corrections and because of this it can be used for TDHF calculations of heavy-ion collisions as well [16, 45]. Pairing is accounted for at the BCS level [46].

The second step is then the calculation of the HF bare nucleus-nucleus potential between two nuclei, also known as the frozen HF potential. This is done by taking two HF ground states for the reaction and putting them in the same calculation box. In this work all frozen HF calculations were made in a box size $67.2 \times 22.4 \times (2 \times 11.2)$ fm^3 in the $x - y - z$ orientation ($z = 0$ is a plane of symmetry) where

[a] e-mail: kirsten.vo-phuoc@anu.edu.au

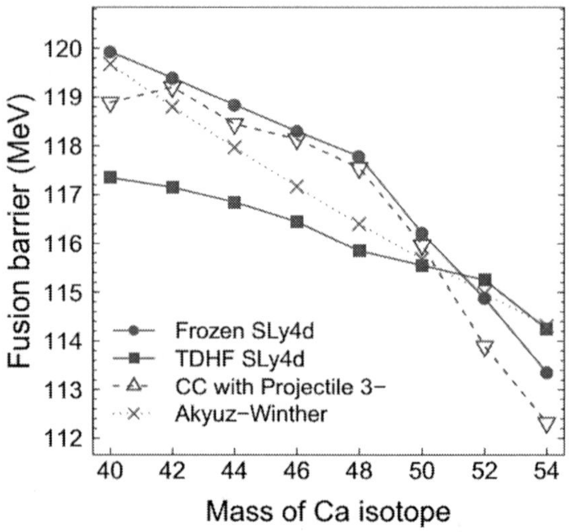

Figure 1. Fusion barriers for ^{A}Ca+^{116}Sn obtained with the frozen HF (circles), TDHF (squares), coupled-channel (triangle) and Akyuz-Winther (crosses) approaches.

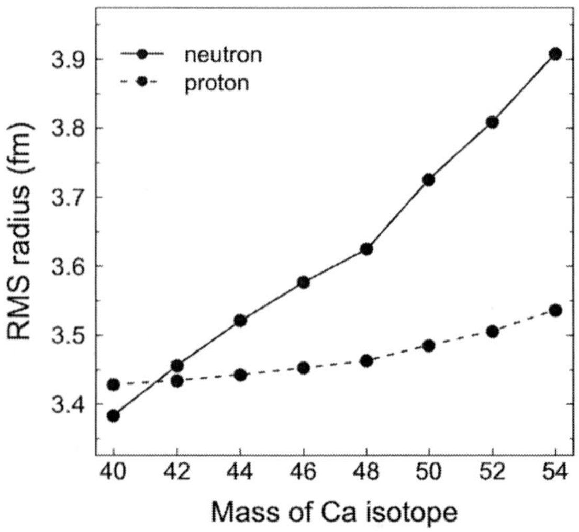

Figure 2. Calcium isotope proton (dashed line) and neutron (solid line) rms radii in the SLy4d parametrisation.

the collision axis is the x–axis. The maximum distance between the centres of masses was 44.8 fm. For every distance between the centres of masses, **r**, the frozen HF potential $V(\mathbf{r})$ is calculated. Using the HF ground state density distributions ρ_1 and ρ_2 of the nuclei, the nuclear part of the potential, $V_N(\mathbf{r})$, is the sum of the energies

$$V_N(\mathbf{r}) = E(\mathbf{r}) - E_{\mathrm{HF}}[\rho_1] - E_{\mathrm{HF}}[\rho_2], \quad (2)$$

for the energy of the total system

$$E(\mathbf{r}) = \int \mathcal{E}[\rho_1(\mathbf{r}') + \rho_2(\mathbf{r} - \mathbf{r}')]d\mathbf{r}', \quad (3)$$

and the (HF) energy for each nucleus

$$E_{\mathrm{HF}}[\rho_j] = \int \mathcal{E}[\rho_j(\mathbf{r})]d\mathbf{r}, \quad (4)$$

where $\mathcal{E}$ is the Skyrme energy density that arises from the Skyrme interaction.

2.2 Static fusion barriers

Fusion barrier energies as obtained by frozen HF are shown in Fig. 1 (circles) alongside the fusion barriers obtained from the Akyüz–Winther potential [47] (crosses). The latter shows a near constant decrease of the fusion barrier energy, V_B, as neutron number is increased. By contrast, the results obtained from static HF calculations show a change of gradient in the V_B line before and after the ^{48}Ca projectile. To explain what causes this trend we can consider some static nuclear properties arising from the HF ground state calculations of the calcium projectiles. One of these properties is the root-mean-square (rms) radius. Both neutron and proton rms radii are shown in Fig. 2. Since the tin target is constant in all reactions, the focus will be on the calcium chain. The neutron rms radius (solid line) exhibits a change of gradient before and after the ^{48}Ca nucleus. An increase in nuclear radius induces a decrease of barrier energies. This is because the point at which the fusion barrier occurs between the nuclei will be at a larger distance thereby decreasing the Coulomb potential and hence the overall fusion barrier. As to why the rms radii behave like this we must turn to single particle shells of the neutrons in the calcium nuclei. The change of gradient happens between the filling of the $1f_{\frac{7}{2}}$ and the $2p_{\frac{3}{2}}$ shells. As ^{48}Ca is a doubly magic nucleus the energy gap between these shells is substantial. The neutrons in the $2p$ shells are less bound, meaning their radial wavefuncion extends to a larger radius than the neutrons in the $1f_{\frac{7}{2}}$ shell which gives rise to a more rapid increase of neutron rms radius.

3 Dynamic effects

3.1 Dynamic fusion barriers

To account for a more realistic picture of fusion reactions, dynamical couplings must be included since nuclei do not necessarily stay in their ground state during fusion reactions. To study this with HF theory, the time-dependent version (TDHF) must be considered. The TDHF equation

$$i\hbar\frac{d\rho}{dt} = [h[\rho], \rho], \quad (5)$$

is solved by the `TDHF3D` code [17] with the `SLy4d` parametrisation of the Skyrme functional as for the static case. A mean field is now defined for the entire system, generated by all the nucleons. The single-particle occupation numbers induced by pairing correlations are kept constant in the dynamics.

To compute a fusion barrier energy for a particular system, as for the static case we again start with both the HF

 DOI: 10.1051/epjconf/201612303001

ground states in the same calculation box at some separation distance. A Galilean boost, $e^{i\mathbf{k}_j\cdot\hat{\mathbf{r}}}$, giving each nucleon in nucleus j momentum is applied to the system at initial time which moves the nuclei towards each other. In TDHF, the fusion probability is either 0 or 1. Here, fusion is deemed successful if a compound nucleus has formed in 5 zs. The TDHF fusion barrier, or "threshold", is found by incrementally increasing the energy for a system until fusion is observed [48].

The TDHF fusion thresholds are shown with squares in Fig. 1. We see that, including dynamics, in most cases, reduces the static fusion barrier. We also observe that including dynamics washes out the static structure effects. Indeed, there is no longer a change of gradient before and after the ^{48}Ca projectile in the TDHF results as seen for the static frozen HF barriers. The expected dynamical effects present in these TDHF calculations are vibrations and transfer. Indeed, coupling to rotational states is not expected to play a major role as the nuclei are essentially spherical. The goal is then to unravel the dynamics in the TDHF calculations to see which dynamical effects have an impact on fusion thresholds.

3.2 Vibrations

Low-lying octupole phonons are known to play an important role in fusion, in particular in reactions with ^{40}Ca [49]. Therefore the coupling to the 3_1^- state is considered for all calcium isotopes. Quadrupole vibrations could also impact fusion. However, the low-lying 2^+ states in ^{40}Ca are not collective. The study of the couplings to 2_1^+ states in the other calcium isotopes will be the subject of future work.

Studies of vibrations of a single nucleus can be done by performing a TDHF calculation for that nucleus in isolation. An octupole boost, $e^{i\varepsilon\hat{Q}_3}$, with boost intensity ε small enough to be in the linear regime and octupole operator

$$\hat{Q}_3 = \sqrt{\frac{7}{16\pi}}\sum_{i=1}^{A}\left(2\hat{z}^3 - 3\hat{z}(\hat{y}^2 + \hat{z}^2)\right), \tag{6}$$

is applied to the HF ground state of calcium isotopes at initial time. The response of the calcium nucleus to this boost is determined by the expectation value of the octupole operator, also known as the octupole moment $\langle\hat{Q}\rangle(t)$.

The `TDHF3D` code is used to compute the evolution of the octupole moment after the initial octupole boost. As the calculations involve only one nucleus, it is possible to use a Skyrme interaction which has been fitted with the center of mass corrections. To study vibrations, we therefore use the `SLy4` paramaterisation [50] which includes such corrections. The octupole moment of the calcium was calculated over 15 zs and is shown in the top panel of Fig. 3 for ^{40}Ca as an example. An oscillation is observed due to the excitation of the octupole phonon.

To resolve an energy spectrum for this, the Fourier transform is taken to produce an octupole strength function, $Q(E)$, as a function of energy (see, e.g., [24]):

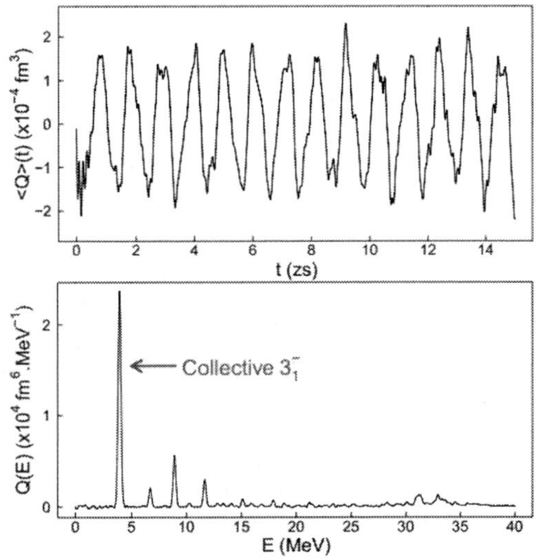

Figure 3. Evolution of the octupole moment in ^{40}Ca (top) and corresponding octupole strength function (bottom).

$$Q(E) = \frac{-1}{\pi\hbar\varepsilon}\int_0^\infty dt\,\langle\hat{Q}\rangle(t)\,\sin(Et/\hbar) \tag{7}$$

$$\equiv \sum_\nu |q_\nu|^2\,\delta(E - E_\nu), \tag{8}$$

where $q_\nu = \langle\nu|\hat{Q}|0\rangle$ is the transition amplitude between the ground state $|0\rangle$ and the excited state $|\nu\rangle$. In practice, $\langle\hat{Q}\rangle(t)$ is multiplied by a damping function to avoid spurious oscillations in the Fourier spectrum due to the finite time of the TDHF calculation. The resulting strength function is shown in the bottom panel of Fig. 3.

The first intense peak of the strength function corresponds to the first collective 3_1^- state. The position of the peak is at the excitation energy of the 3_1^- state (as seen by TDHF) and the area under the peak corresponds to the transition probability $|q_3|^2$. This can be used to determine the deformation parameter β_3 according to [51]

$$\beta_3 = \frac{4\pi|q_3|}{3R_0^3A}, \tag{9}$$

where R_0 is the nuclear radius. The energy and deformation parameter of the 3_1^- state in ^{40}Ca obtained with TDHF are compared with experimental values in Table 1. Despite an underestimation of the deformation parameter, the agreement between theory and experiment is reasonable. Similar calculations of 3_1^- properties have been computed with TDHF for the other calcium isotopes.

To see what this one dynamical effect has on the fusion barriers of the systems, the CC approach [52], using the `CCFULL` code [53] is used since dynamical effects can be added in one at a time. Couplings to low-lying states could induce a complex structure in the distribution of fusion barriers [2]. The main feature, however, is a lowering of the main peak in the barrier distribution, often resulting

	$E_x(3_1^-)$ (MeV)	β_3
Exp.	3.737	0.3-0.4
TDHF	3.916	0.184

Table 1. Comparison of excitation energy and deformation parameter between TDHF and experimental values [51].

in a lowering of the average barrier. These average barriers, calculated with coupling to the 3_1^- state in the calcium isotopes, are shown in Fig. 1. As expected, this coupling lowers the barrier, in comparison to the frozen HF static barrier, in all cases. Note that couplings to low-lying 2^+ states, not included here, are expected to produce a similar effect.

The inclusion of couplings to low-lying vibrations is helpful for explaining the difference between the static and dynamic cases for the systems with $^{40-50}$Ca as projectiles but not for the two most neutron-rich systems. Indeed, we see in Fig. 1 that for the ^{52,54}Ca+^{116}Sn systems, including dynamics actually increased the fusion barrier as the TDHF fusion threshold is higher than the frozen HF bare potential barrier. Therefore, another dynamical effect aside from vibrations must be considered to help explain this observation.

3.3 Transfer

Nucleon transfer in reactions acts to balance out the N/Z ratio between the target and projectile. This effect is accounted for in TDHF calculations [33, 54]. The N/Z ratio of ^{116}Sn is 1.32 and the N/Z ratio for the calcium projectiles range from 1 to 1.70. For this proceeding, the focus is on proton transfer for these systems. Given these N/Z values, it would be expected from this preliminary argument that proton transfer would happen in the direction from projectile to target for the lighter systems, and from target to projectile in heavier systems.

Nucleon transfer is incorporated into TDHF calculations and is observed during TDHF reactions of the systems $^{40-54}$Ca+^{116}Sn at centre of mass energies 99% of the fusion threshold for both protons and neutrons. Proton transfer probabilities are calculated using the particle number projection method [31]. The probabilities for the different Z number of the target-like fragment for the systems ^{40,54}Ca+^{116}Sn can be seen in Fig. 4.

For ^{40}Ca+^{116}Sn, there is a large probability for finding $Z = 51$ in the target-like fragment. This corresponds to a likelihood of one proton transfer from projectile to target. By contrast, proton transfer in the opposite direction is likely to occur for ^{54}Ca+^{116}Sn. This is consistent with what is seen in the TDHF fusion barrier thresholds. Proton transfer in either direction will change the Coulomb repulsion. In particular, one proton transfer from ^{54}Ca to ^{116}Sn increases the Coulomb potential, $V_C = e^2 Z_1 Z_2 / r$ by 2.7% which in turn increases the fusion barrier. This is a possible explanation for why the dynamical TDHF fusion threshold is higher than the static frozen HF barrier for this system.

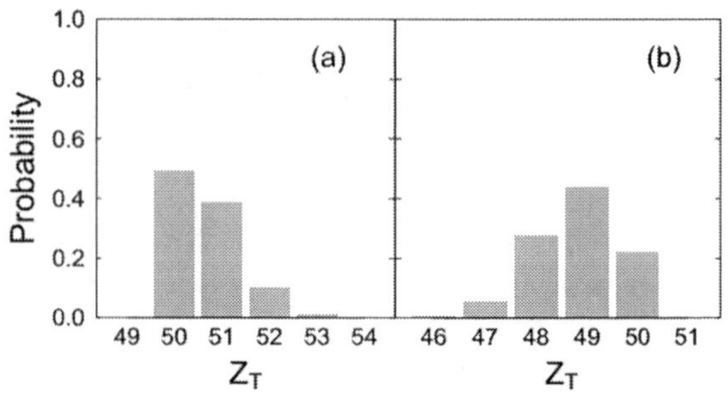

Figure 4. TDHF proton number probabilities for the target-like fragment in the exit channel of (a) ^{40}Ca+^{116}Sn and (b) ^{54}Ca+^{116}Sn reactions at $E_{cm} = 0.99 V_B$.

4 Conclusions

Microscopic theory provides unique insight into nuclear structure and subsequently the effects of structure on reactions. In the static picture of fusion, the HF ground state properties such as rms neutron radius of the calcium isotopes have an effect on the fusion barrier. Nuclei with less bound neutrons and an increasing neutron skin (calcium isotopes heavier than ^{48}Ca) will lead to a more rapid decrease of the fusion barrier. When dynamics are included into reactions, it was seen that for most of the cases this decreased the fusion barrier. However, it increased the fusion barrier for the two most neutron rich systems. Including dynamics into the calculations overrode the static structure effects completely. Coupled-Channels analysis showed that coupling to the 3_1^- state of the calcium projectiles lowered the fusion barrier in all cases. Considering other dynamical effects such as proton transfer is important for the most neutron rich systems. Future works will consider the effect of couplings to low-lying quadrupole vibrations as well as the impact of neutron transfer on fusion.

Acknowledgements

This work has been supported by the Australian Research Council under Grant No. FT120100760. The calculations in this proceedings were done with the assistance of resources from the National Computational Infrastructure (NCI) which is supported by the Australian Government.

References

[1] N. Rowley, G. R. Satchler and P. H. Stelson, Phys. Lett. B **254**, 25 (1991).
[2] M. Dasgupta, D. J. Hinde, N. Rowley *et al.*, Annu. Rev. Nucl. Part. Sci. **48**, 401 (1998).
[3] C. R. Morton, M. Dasgupta, D. J. Hinde *et al.*, Phys. Rev. Lett. **72**, 4074 (1994).
[4] J. R. Leigh, M. Dasgupta, D. J. Hinde *et al.*, Phys. Rev. C **52**, 3151 (1995).
[5] A. M. Stefanini, D. Ackermann, L. Corradi *et al.*, Phys. Rev. Lett. **74**, 864 (1995).

[6] M. Beckerman, M. Salomaa, A. Sperduto *et al.*, Phys. Rev. Lett. **45**, 1472 (1980).
[7] A. Stefanini, G. Fortuna, R. Pengo *et al.*, Nucl. Phys. A **456**, 509 (1986).
[8] H. Timmers, L. Corradi, A. Stefanini *et al.*, Phys. Lett. B **399**, 35 (1997).
[9] C. L. Jiang, K. E. Rehm, B. B. Back *et al.*, Phys. Rev. C **89**, 051603 (2014).
[10] C. L. Jiang, A. M. Stefanini, H. Esbensen *et al.*, Phys. Rev. C **91**, 044602 (2015).
[11] M. Dasgupta, D. J. Hinde, R. D. Butt *et al.*, Phys. Rev. Lett. **82**, 1395 (1999).
[12] A. Chatterjee, A. Navin, A. Shrivastava *et al.*, Phys. Rev. Lett. **101**, 032701 (2008).
[13] A. Lemasson, A. Navin, N. Keeley *et al.*, Phys. Rev. C **82**, 044617 (2010).
[14] A. Lemasson, A. Navin, M. Rejmund *et al.*, Phys. Lett. B **697**, 454 (2011).
[15] J. W. Negele, Rev. Mod. Phys. **54**, 913 (1982).
[16] C. Simenel, Eur. Phys. J. A **48**, 152 (2012).
[17] K. H. Kim, T. Otsuka and P. Bonche, J. Phys. G: Nucl. Part. Phys. **23**, 1267 (1997).
[18] A. S. Umar and V. E. Oberacker, Phys. Rev. C **71**, 034314 (2005).
[19] T. Nakatsukasa and K. Yabana, Phys. Rev. C **71**, 024301 (2005).
[20] J. A. Maruhn, P. G. Reinhard, P. D. Stevenson *et al.*, Phys. Rev. C **71**, 064328 (2005).
[21] C. Simenel and P. Chomaz, Phys. Rev. C **80**, 064309 (2009).
[22] I. Stetcu, A. Bulgac, P. Magierski *et al.*, Phys. Rev. C **84**, 051309 (2011).
[23] S. Fracasso, E. B. Suckling and P. D. Stevenson, Phys. Rev. C **86**, 044303 (2012).
[24] B. Avez and C. Simenel, Eur. Phys. J. A **49**, 76 (2013).
[25] G. Scamps and D. Lacroix, Phys. Rev. C **88**, 044310 (2013).
[26] C. Simenel, R. Keser, A. S. Umar *et al.*, Phys. Rev. C **88**, 024617 (2013).
[27] C. Simenel, M. Dasgupta, D. J. Hinde *et al.*, Phys. Rev. C **88**, 064604 (2013).
[28] A. S. Umar, C. Simenel and V. E. Oberacker, Phys. Rev. C **89**, 034611 (2014).
[29] G. F. Dai, L. Guo, E. G. Zhao *et al.*, Phys. Rev. C **90**, 044609 (2014).
[30] K. Washiyama, Phys. Rev. C **91**, 064607 (2015).
[31] C. Simenel, Phys. Rev. Lett. **105**, 192701 (2010).
[32] C. Simenel, Phys. Rev. Lett. **106**, 112502 (2011).
[33] K. Sekizawa and K. Yabana, Phys. Rev. C **88**, 014614 (2013).
[34] K. Sekizawa and K. Yabana, Phys. Rev. C **90**, 064614 (2014).
[35] G. Scamps and D. Lacroix, Phys. Rev. C **87**, 014605 (2013).
[36] A. Wakhle, C. Simenel, D. J. Hinde *et al.*, Phys. Rev. Lett. **113**, 182502 (2014).
[37] V. E. Oberacker, A. S. Umar and C. Simenel, Phys. Rev. C **90**, 054605 (2014).
[38] A. S. Umar, V. E. Oberacker and C. Simenel, Phys. Rev. C **92**, 024621 (2015).
[39] C. Simenel and A. S. Umar, Phys. Rev. C **89**, 031601 (2014).
[40] G. Scamps, C. Simenel and D. Lacroix, Phys. Rev. C **92**, 011602 (2015).
[41] P. Goddard, P. Stevenson and A. Rios, Phys. Rev. C **92**, 054610 (2015).
[42] P. Goddard, P. Stevenson and A. Rios, Phys. Rev. C **93**, 014620 (2016).
[43] P. Bonche, H. Flocard and P. H. Heenen, Comp. Phys. Com. **171**, 49 (2005).
[44] T. Skyrme, Phil. Mag. **1**, 1043 (1956).
[45] M. Kortelainen, J. McDonnell, W. Nazarewicz *et al.*, Phys. Rev. C **85**, 024304 (2012).
[46] K. Bennaceur, P. Bonche and J. Meyer, C. R. Physique **4**, 555 (2003).
[47] R. A. Broglia and A. Winther, *Heavy Ion Reaction Lecture Notes, Vol. I: Elastic and Inelastic Reactions* (Benjamin/Cummings, Reading, MA, 1981)
[48] C. Simenel and B. Avez, Int. J. Mod. Phys. E **17**, 31 (2008).
[49] K. Hagino and N. Takigawa, Phys. Rev. Lett. **79**, 2014 (1997).
[50] E. Chabanat, P. Bonche, P. Haensel *et al.*, Nucl. Phys. A **635**, 231 (1998).
[51] T. Kibedi and R. H. Spear, Atomic Nuclear and Data Tables **80**, 35 (2002).
[52] K. Hagino and N. Takigawa, Prog. Th. Phys. **128**, 1061 (2012).
[53] K. Hagino, N. Rowley and A. T. Kruppa, Comp. Phys. Com. **123**, 143 (1999).
[54] C. Simenel, D. J. Hinde, R. du Rietz *et al.*, Phys. Lett. B **710**, 607 (2012).

EPJ Web of Conferences **123**, 03002 (2016)
Heavy Ion Accelerator Symposium 2015 DOI: 10.1051/epjconf/201612303002

Resonances in transfer-triggered breakup of ^{7}Li in near-barrier collisions

E. C. Simpson[1,a], K. J. Cook[1], M. Dasgupta[1], S. Kalkal[1], D. H. Luong[1], I. P. Carter[1], D. J. Hinde[1], and E. Williams[1]

[1] *Department of Nuclear Physics, Research School of Physics and Engineering, The Australian National University, Canberra, ACT 2601, Australia*

Abstract. Above-barrier complete fusion cross sections of weakly-bound ^{6,7}Li and ^{9}Be are known to be suppressed with respect to single-barrier penetration model calculations. Breakup of the projectile — either via direct excitation of continuum states, or by transfer of nucleons — is thought to be the cause, preventing complete capture of the projectile charge. Using the example of ^{7}Li→^{8}Be→ $\alpha + \alpha$ we show how the contributions to breakup from different resonances in ^{8}Be can be identified, and discuss their likely influence on fusion.

1 Breakup and fusion suppression

Above-barrier fusion of light, weakly-bound projectiles such as ^{7}Li is known to be suppressed [1–3]: single-barrier penetration model calculations, which accurately model fusion of strongly-bound projectiles such as ^{18}O, overestimate measured fusion cross sections for weakly bound ^{7}Li by ~25% (see Ref. [4] for a recent review). Breakup reactions are thought to reduce the probability for fusion of the entire projectile. Both direct breakup (^{7}Li→ $\alpha + t$) and transfer-induced breakup (e.g., $^7\text{Li}\xrightarrow{+p}{}^8\text{Be}\rightarrow \alpha + \alpha$, $^7\text{Li}\xrightarrow{-n}{}^6\text{Li}\rightarrow \alpha + d$, $^7\text{Li}\xrightarrow{-2n}{}^5\text{Li}\rightarrow \alpha + p$) have been found to be important with ^{208}Pb and ^{209}Bi targets [5, 6].

Understanding the detail of these breakup processes is crucial to understand their influence on fusion. Breakup can only suppress complete fusion when the disintegration occurs when the projectile and target nucleus are approaching one another. However, from below-barrier measurements of breakup, where incomplete fusion is negligible, it is known that breakup often proceeds via long-lived resonance states [2]. These states, such as the ^{8}Be ground state, survive until the ejectile is very far from the target, and *cannot* suppress fusion. Short-lived states will decay much nearer the target, but even sub-zeptosecond ($<10^{-21}$ s) lifetimes may be sufficient to alter whether the projectile-like nucleus breaks up prior to reaching the fusion barrier, and therefore reduce the impact of breakup on fusion.

To investigate these effects we consider the example of proton pickup by ^{7}Li on a ^{58}Ni target, giving ^{8}Be. ^{8}Be is unbound with respect to decay into $\alpha + \alpha$, with the narrow 0^+ ground state 0.092 MeV above the threshold, and a broad 2^+ resonance at 3.12 MeV [7]. The ^{58}Ni target is chosen to reduce, with respect to heavier ^{208}Pb-region targets, the influence that the target has on the trajectories of the fragments following breakup. Identification of asymptotic and near-target breakup, classical dynamical simulations of the resultant α fragment trajectories, and their respective influence on fusion is discussed below.

2 Identifying asymptotic breakup

When investigating the impact of breakup on fusion, the most important characteristic of the reaction is where (and when) the breakup occurs. Only if the projectile breaks up prior to reaching the fusion barrier will it affect fusion. At above barrier energies, it is likely that one or other of the fragments may still fuse with the target, giving a contribution to incomplete fusion, and making it harder to study the breakup itself. A series of measurements at below barrier energies has been made at the Australian National University Heavy Ion Accelerator Facility to clarify the detail of the breakup mechanism. The measurements used the BALiN double-sided silicon-strip detector array [8] to detect the energies E_i and angles (θ_i, ϕ_i) of the resulting pairs of charged fragments detected in coincidence.

The relative energy of the fragments $E_{rel} = \frac{1}{2}\mu v_{12}^2$ gives a basic way to distinguish between breakup of long-lived states and that occurring near the target nucleus. Long-lived states have a well defined energy that is unperturbed by any differential acceleration of the fragments by the (highly remote) target following breakup. As such, when the relative energy of the fragments is reconstructed, these states appear as narrow peaks in the E_{rel} spectrum. The example of $^7\text{Li}\xrightarrow{+p}{}^8\text{Be}\rightarrow \alpha\text{-}\alpha$ is shown in Fig. 1, for a ^{58}Ni target with a beam energy of 13.1 MeV (~ $0.95V_B$). The data is gated on the 2.981-MeV $1/2^+$ state in ^{57}Co, which is strongly populated in other proton pickup reactions with ^{7}Li [9]. The red shaded region highlights the narrow peak associated with the ^{8}Be ground state, which sits on an exponentially decaying background.

[a]e-mail: edward.simpson@anu.edu.au

DOI: 10.1051/epjconf/201612303002

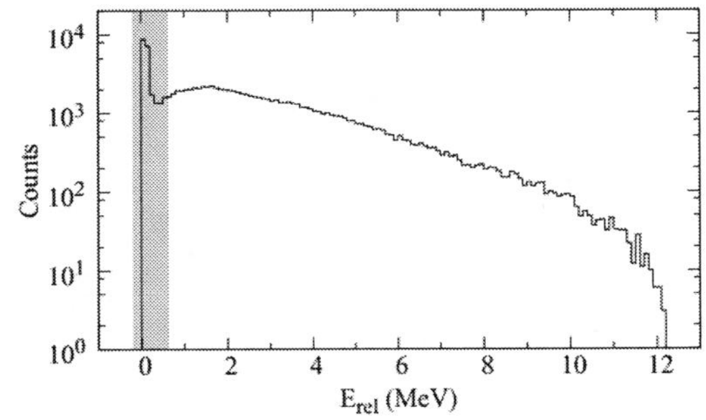

Figure 1. The relative energy of the two fragments can be used to separate near-target and asymptotic breakup. Shown are the measured α-α relative energies for $^{7}Li+^{58}Ni$ collisions. Long-lived states, such as the ^{8}Be ground state highlighted by the red area, decay very far from the target, and as such, the relative energy of the αs remains equal to the energy of the state. The exponentially decaying component of the distribution is assumed to result from breakup near the target.

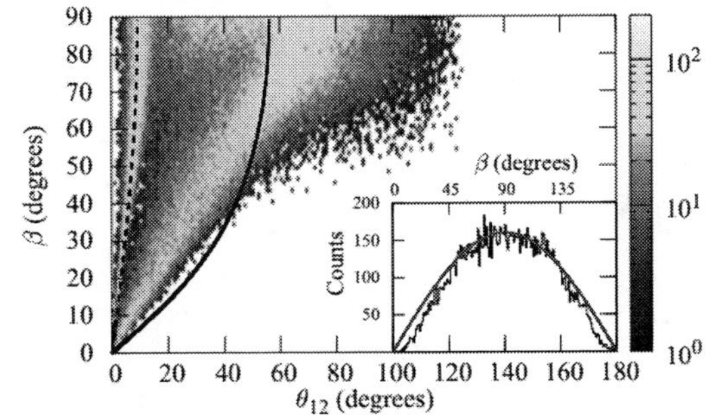

Figure 2. The distribution of data with respect to the angles θ_{12} and β, shown here for $^{7}Li+^{58}Ni$. The dashed and solid lines indicate the expected correlation for asymptotic decay of the ^{8}Be 0^{+} and 2^{+} states respectively. The 0^{+} band is clearly asymptotic, whereas the 2^{+} is not. The inset shows the projection of the ground state band onto β, which is consistent with isotropic emission ($\sin\beta$ curve shown in red).

3 Near-target breakup

Whilst asymptotic breakup manifests as a narrow peak in the relative energy distribution, the origin of the remainder — the featureless, exponential background — is less clear. It has previously been assumed [5, 6] that it arises from breakup near the target nucleus. Half of these events were assumed to occur as the reaction partners approached one another, therefore half were supposed capable of suppressing fusion. For the case of $^{9}Be\rightarrow{}^{8}Be\rightarrow\alpha+\alpha$, it was suggested that the near-target breakup might proceed via instantaneous breakup of the ^{8}Be 2^{+} state [10]. Were it not for the presence of the target, the 2^{+} would give a peak at E_{rel} = 3.12 MeV.

A more refined understanding is desirable. Useful insights can be obtained by studying the distribution of events as a function of the fragment opening angle θ_{12}, and the orientation of the fragments' relative motion with respect to that of their centre of mass, β [11, 12]. The opening angle θ_{12} is typically strongly correlated with the relative energy E_{rel} (see e.g., Fig. 8(a) of Ref. [6]). If the relative motion is aligned with the motion of their centre-of-mass in the laboratory ($\beta = 0°$ and $\beta = 180°$), then the opening angle θ_{12} will be minimised. Conversely, if the relative motion is perpendicular to that of the centre-of-mass ($\beta = 90°$), the θ_{12} will be maximised and, for the case of fragments of identical mass, the energies of the fragments E_1 and E_2 will be equal. If there is no acceleration of the α fragments by the target following breakup, there is a well defined relationship between the orientation angle β and the opening angle of the two fragments θ_{12} for a given relative energy E_{rel} (the asymptotic correlation).

Differential acceleration of the fragments following breakup distorts this relationship. The experimental data are shown in Fig. 2 compared to the expectations for asymptotic breakup of the ^{8}Be 0^{+} (E_{rel} = 0.092 MeV) and 2^{+} (E_{rel} = 3.12 MeV) states. Since the two fragments are identical, the data is symmetric about $\beta = 90°$ and has been folded about this point. The data is limited to $\theta_{12} < 132°$ — with limited efficiency for the largest θ_{12} — due to the incomplete coverage of the BALIN detector array.

Two bands are seen in the data. The band on the left is associated with breakup via the ^{8}Be 0^{+} ground state, and clearly follows the expected asymptotic correlation (dashed line). Projected onto β (inset of Fig. 2), the sinusoidal distribution for this band indicates isotropic breakup. The second band does not follow the asymptotic expectation, but it is interesting that it arises from the almost featureless relative energy distribution. The relative closeness to the asymptotic prediction for the ^{8}Be 2^{+} (solid line) suggests these events do indeed originate from this state.

4 Modelling post-breakup acceleration

To investigate how the distribution of events with β and θ_{12} is sensitive to the location of breakup, we need to model the trajectories of the fragments after breakup. Assuming that near-target breakup originates from ^{8}Be in the 2^{+} resonant state, the initial relative energy of the fragments will be equal to the resonance energy, E_R = 3.12 MeV [7]. How the α particles accelerate will depend on (a) their orientation with respect to the target [13], and (b) their distance from the target. This is illustrated in Fig. 3.

To investigate this further we make Monte Carlo simulations using a modified version of the PLATYPUS code [14, 15] to model post-breakup acceleration effects. The breakup location is fixed at some point Δr on the projectile-target trajectory, where here Δr is some radial distance after the distance of closest approach R_0. This is illustrated in Fig. 4. The fragments are initially placed with their centre of mass at the breakup location, and are separated by a distance equal to their mutual barrier radius (~ 6 fm). Breakup is thus defined to occur when the two fragments have passed their mutual barrier. Prior to this

DOI: 10.1051/epjconf/201612303002

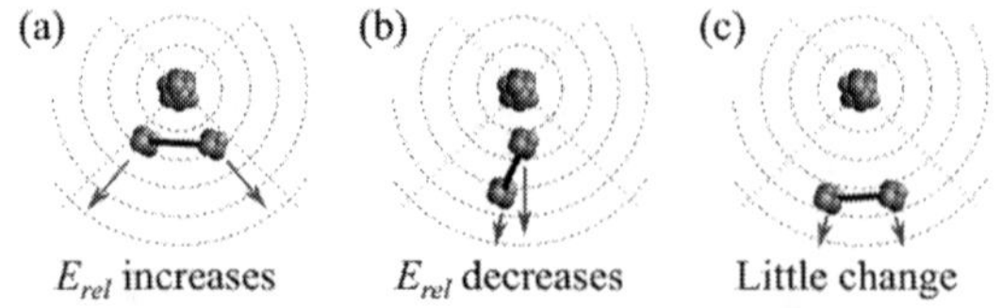

Figure 3. The acceleration of the fragments, following breakup, depends on their initial orientation with respect to the target nucleus. (a) If the relative vector of the α-particles is aligned perpendicular to the field of the target, the post-breakup acceleration will increase the α-α relative energy. (b) Conversely, alignment with the field will tend to reduce the relative energy. (c) The magnitude of the acceleration depends on the proximity to the target - the further from the target, the smaller the post-breakup acceleration.

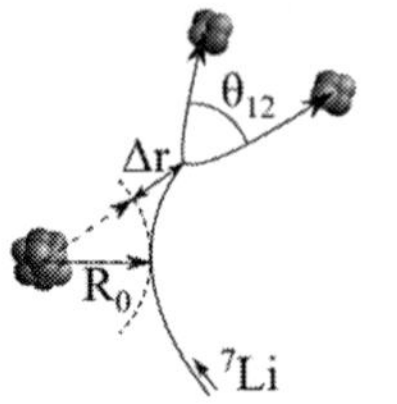

Figure 4. Illustration of the projectile-target trajectory. In the simulations, breakup is assumed to occur at some fixed radial distance Δr after the distance of closest approach R_0 has been passed. The opening angle of the fragment asymptotic velocity vectors is θ_{12}.

point, we assume that the fragments remain sufficiently well localised so as not to affect fusion. Tidal forces generated by the target may be important in hastening or indeed hindering breakup, but this is beyond the scope of the present discussion. The orientation of the α-α relative position vector is randomly sampled from an isotropic distribution. Having set these initial conditions, the fragments and target propagate according to their mutual Coulomb interactions, and the simulation is continued until all three are sufficiently far apart. Five hundred events are sampled to construct the correlation between asymptotic quantities β and θ_{12}. Further details can be found in Ref. [12].

5 Results for ^{7}Li $\rightarrow$ ^{8}Be $\rightarrow \alpha + \alpha$

Here we focus on events when the projectile-like and target-like nuclei are receding from each other. Breakup prior to the projectile reaching the distance of closest approach, and the effects of resonance width and projectile-target angular momentum, are discussed in Ref. [12]. The present calculations are for zero projectile-target angular momentum, and this axial symmetry means that for each calculation all events lie along a single line. The results for distances Δr = 1, 2, 4, 10 and 1000 fm after the distance of closet approach are shown in Fig. 5. At 1000 fm the breakup is essentially asymptotic, and the correlation found agrees with the analytic expectation shown in Fig. 2. As the breakup point moves inwards towards the target nucleus, the correlation becomes distorted. For those events where the αs are aligned with the target potential [e.g., Fig. 3(b)], the acceleration reduces the relative energy and θ_{12}. For the α particles aligned perpendicular to the potential, the acceleration increases the relative energy and θ_{12}.

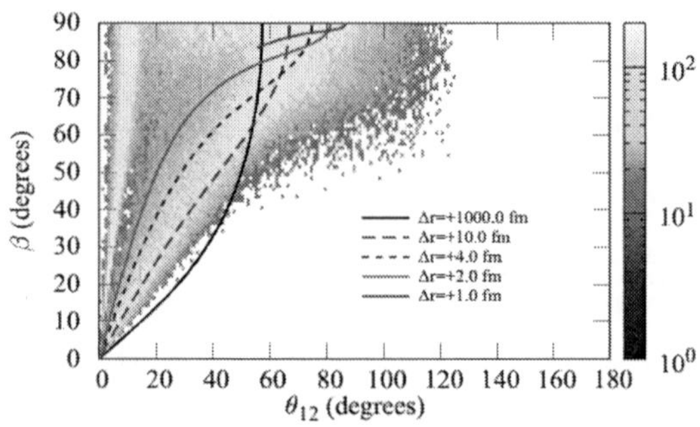

Figure 5. Data and simulations showing the correlation between β and θ_{12}. See the text for further discussion.

As the distance becomes smaller the distortions become stronger. For the smallest distance, Δr = 1 fm, nearly all the energy is locked into the fragment-target potential. The release of this energy, in essentially independent interactions of the fragments with the target, gives fragments with very similar energies. As a result, all 500 events reconstruct to β values near 90°.

The results of the simulations suggest that much of what is seen in the present experiment arises from breakup occurring only once the projectile and target have receded by several femtometers from the distance of closest approach. Breakup when the two are approaching tends to lead to large θ_{12} which the current experiment is not sensitive to — further details can be found in Ref. [12].

6 Consequences for fusion

These results suggest that the properties of short-lived resonances [16, 17], such as the ^{8}Be 2^+, may play an important role in determining the extent to which breakup suppresses fusion. This is illustrated in Fig. 6. The plot shows the radial separation of the projectile and target as a function of time for a sample trajectory, coloured by the anticipated transfer probability. This transfer probability is expected to fall exponentially with increasing separation (see e.g., [14, 18]). The transfer probability is peaked at the distance of closest approach, but if the state populated by the transfer has a lifetime of 0.5 zs, the breakup locations will be shifted (approximately) along the trajectory to later times (shown in green). This lifetime effect could explain why the simulations suggest breakup occurs primarily when the projectile and target are already moving apart. Moreover, at above barrier energies the reactants will have crossed the fusion barrier before the resonant state has had a chance to disintegrate, and so fusion will not be suppressed.

7 Conclusions and outlook

The correlation between β and θ_{12} appears to be sensitive to the distance between the projectile-like and target-like nuclei at the point of breakup. The properties of broad

DOI: 10.1051/epjconf/201612303002

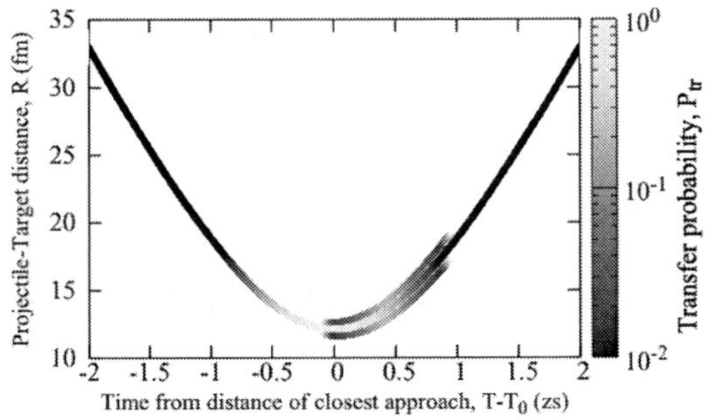

Figure 6. Illustration of the influence of short-lived states on breakup. The yellow-red-black line shows the projectile-target trajectory as a function of the time from the distance of closest approach and projectile-target separation R. It is coloured by the transfer probability, which peaks at the distance of closest approach and falls exponentially, as show in the inset. The green region indicates the approximate breakup location, shifted due to the unbound state populated having a short lifetime (~ 0.5 zs).

resonant states may be very important in determining this distance, and therefore the consequences for fusion. To further understand the influence of breakup on fusion, new measurements are required on a variety of targets to focus on large opening angles θ_{12}, to study breakup occurring prior to the distance of closest approach. More sophisticated classical dynamical modelling must also be performed, taking into account both narrow and broad resonances and their characteristic shapes and lifetimes.

Furthermore, every nucleus, particularly every light nucleus, is different. Due to differences in their structure the results found for β vs. θ_{12} for ^{6}Li→ $\alpha + d$ and ^{7}Li→ $\alpha + t$ coincidences are very different from those shown here — both have long-lived narrow states, and though they show evidence of near-target breakup, neither has low-lying broad resonant states similar to the ^{8}Be 2^+. The consequences for fusion may therefore be rather different, and require further detailed investigations.

Acknowledgements

Helpful discussions with A. Diaz-Torres are gratefully acknowledged. We thank M. Evers, L. R. Gasques, P. R. S. Gomes, R. Linares and A. Wakhle for their assistance in running the experiment. Support from Australian Research Council grants FL110100098, DP130101569, DE140100784 and DP14101337 is acknowledged. Support for accelerator operations through the NCRIS program is acknowledged.

References

[1] M. Dasgupta, D. J. Hinde, K. Hagino *et al.*, Phys. Rev. C **66**, 041602(R) (2002).
[2] D. J. Hinde, M. Dasgupta, B. R. Fulton *et al.*, Phys. Rev. Lett. **89**, 272701 (2002).
[3] M. Dasgupta, P. R. S. Gomes, D. J. Hinde *et al.*, Phys. Rev. C **70**, 024606 (2004).
[4] L. F. Canto, P. R. S. Gomes, R. Donangelo *et al.*, Phys. Rep. **596**, 105 (2015).
[5] D. H. Luong, M. Dasgupta, D. J. Hinde *et al.*, Phys. Lett. B **695**, 105 (2011).
[6] D. H. Luong, M. Dasgupta, D. J. Hinde *et al.*, Phys. Rev. C **88**, 034609 (2013).
[7] D. R. Tilley, J. H. Kelley, J. L. Godwin *et al.*, Nucl. Phys. **A745**, 155 (2004).
[8] D. H. Luong, K. J. Cook, E. Williams *et al.*, Proc. of Sci. **Pos(NIC XII)**, 185 (2012).
[9] A. Marinov, W. Oelert, S. Gopal *et al.*, Nucl. Phys. **A438**, 429 (1985).
[10] R. Rafiei, R. du Rietz, D. H. Luong *et al.*, Phys. Rev. C **81**, 024601 (2010).
[11] A. B. McIntosh, S. Hudan, C. J. Metelko *et al.*, Phys. Rev. Lett. **99**, 132701 (2007).
[12] E. C. Simpson, K. J. Cook, D. H. Luong *et al.*, Phys. Rev. C **93**, 024605 (2016).
[13] J. E. Mason, S. B. Gazes, R. B. Roberts *et al.*, Phys. Rev. C **45**, 2870 (1992).
[14] A. Diaz-Torres, D. J. Hinde, J. A. Tostevin *et al.*, Phys. Rev. Lett. **98**, 152701 (2007).
[15] A. Diaz-Torres, Comp. Phys. Comm. **182**, 1100 (2011).
[16] A. M. Lane and R. G. Thomas, Rev. Mod. Phys. **30**, 257 (1958).
[17] F. C. Barker, Aust. Jour. Phys. **41**, 743 (1988).
[18] S. Saha, Y. K. Agarwal and C. V. K. Baba, Phys. Rev. C **49**, 2578 (1994).

EPJ Web of Conferences **123**, 03003 (2016)
Heavy Ion Accelerator Symposium 2015
DOI: 10.1051/epjconf/201612303003

Direct reactions for nuclear structure required for fundamental symmetry tests

P. E. Garrett[1,a], E. T. Rand[1], A. Diaz Varela[1], G.C. Ball[2], V. Bildstein[1], T. Faestermann[3], B. Hadinia[1], R. Hertenberger[4], D. S. Jamieson[1], B. Jigmeddorj[1], K. G. Leach[1,5], C. E. Svensson[1], and H.-F. Wirth[4]

[1] *Department of Physics, University of Guelph, Guelph, ON, N1G2W1 Canada*
[2] *TRIUMF, 4004 Wesbrook Mall, Vancouver, BC V6T2A3 Canada*
[3] *Physik Department, Technische Universität München, D-85748 Garching, Germany*
[4] *Fakultät für Physik, Ludwig-Maximilians-Universität München, D-85748 Garching, Germany*
[5] *Department of Physics, Colorado School of Mines, Golden CO, 80401 USA*

Abstract. A program of nuclear structure studies to support fundamental symmetry tests has been initiated. Motivated by the search for an electric dipole moment in ^{199}Hg, the structure in the vicinity has been explored via direct reaction studies. To date, these have included the ^{198,200}Hg(d,d') inelastic scattering reactions, with the aim to obtain information on the $E2$ and $E3$ strength distributions, and the ^{198}Hg(d,p) and ^{200}Hg(d,t) reactions to obtain information on the single-particle states in ^{199}Hg. The studies using the ^{200}Hg targets have been fully analyzed using the FRESCO reaction code yielding the $E2$ and $E3$ strength distribution to 4 MeV in excitation energy, and the (d,t) single- particle strength to over 3 MeV in excitation energy.

1 Introduction

The combination of charge-conjugation (C) and parity (P) violating processes is essential for the explanation of the matter-antimatter asymmetry observed in the Universe [1]. The CPT Theorem [2] states that invariance under transformations of the combination of the three fundamental symmetries, C, P, and time-reversal transformation (T) is strictly obeyed. To date, all experimental evidence supports that CPT is a true symmetry of nature, and thus a violation of CP, as required for baryogenesis, implies T violation. A particle electric dipole moment (EDM) is odd under both the P and T transformations, therefore a permanent EDM for an elementary particle or atom can only arise from parity *and* time-reversal violating fundamental interactions.

The Standard Model sources of CP violation, in the weak interaction represented by the complex phase δ_{CKM} of the Cabibbo-Kobayashi-Maskawa (CKM) matrix, and in a phase that arises from the vacuum expectation value of the quantum chromodynamics (QCD) gluon field δ_{QCD}, are insufficient to account for the matter-antimatter asymmetry. Additional sources of CP violation are therefore required, providing a strong motivation to search for new CP-violating physics beyond the Standard Model. Present theories beyond the Standard Model, such as multiple-Higgs theories, left-right symmetric, and supersymmetry (SUSY), generally predict EDMs within current experimental reach [3]. Furthermore, present upper limits on the EDMs or the neutron, electron and ^{199}Hg atom have already significantly reduced the parameter spaces of these models. Currently, the upper limit on the EDM of the ^{199}Hg atom provides the most stringent limit on many possible CP-violating terms [4]. Measuring an EDM in a neutral atom is complicated by orbiting atomic electrons, which would arrange themselves to exactly cancel an EDM if the nucleus were a point-like object. Fortunately, nuclei have a finite size and the screening effect does not completely cancel the observable atomic EDM. The intrinsic Schiff moment, the lowest order time-reversal odd moment of a nucleus that is measurable in a neutral atom [5], is a measure of the difference between the charge and dipole distributions of the nucleus. It is responsible for inducing the observable atomic EDM in the electron cloud of the atom. To gain a better understanding the distribution of Schiff strength in ^{199}Hg, higher precision data on the $E1$, $E2$, and $E3$ strength distributions are required.

With the aim of providing greater knowledge of the electric strength distributions in the Hg isotopes than is currently available, a program of study of the inelastic scattering reactions has been initiated, with ^{200}Hg the first nucleus studied. Inelastic scattering provides information on nuclear matrix elements, including those for high multipolarities that typically cannot be obtained in Coulomb excitation experiments employing γ-ray detection. Additionally, nucleon transfer reactions are being re-investigated with the aim of achieving higher sensitivity and information on states at higher excitation energy.

The Hg experimental program represents an expansion of an on-going program of direct reaction studies moti-

[a] e-mail: pgarrett@physics.uoguelph.ca

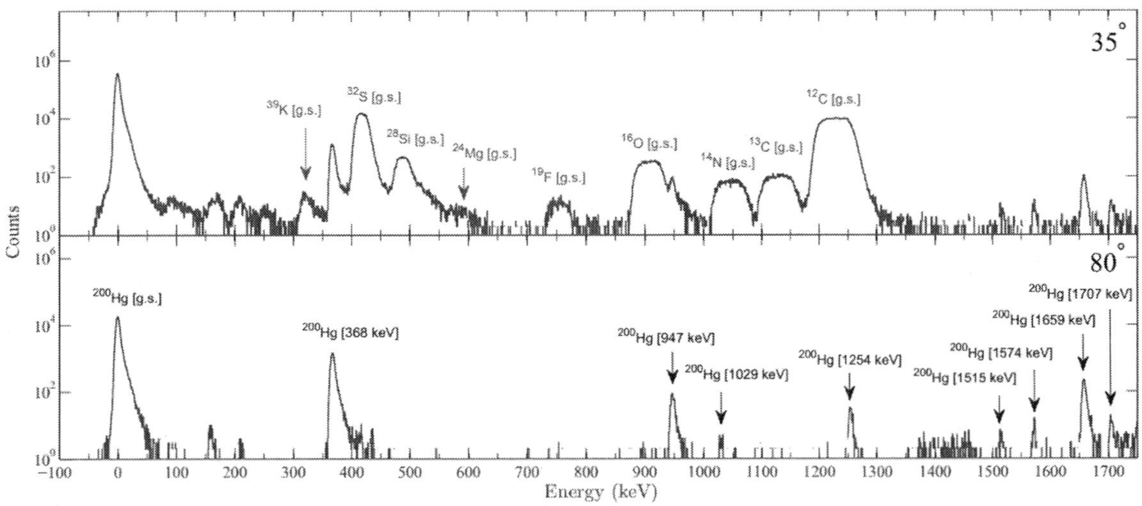

Figure 1. Portion of the spectrum obtained at 35° (top) and 80° (bottom) from the bombardment of the ^{200}Hg^{32}S target with 22 MeV deuterons. The deuterons originating from the light-mass impurities, having different kinematics and focusing properties than those from the ^{200}Hg, can be identified by their width and movement as a function of angle with respect to the ^{200}Hg peaks. This high-energy portion of the deuteron spectrum has many of the light-mass impurity labelled by their origin in the top panel; at high scattering angles these light-mass impurity peaks are no longer present in the spectrum which reveals a number of lower-intensity ^{200}Hg peaks from inelastic scattering.

vated by fundamental symmetry tests, especially daughter nuclei of superallowed Fermi β emitters. For example, one- and two-neutron transfer measurements on ^{52}Cr and ^{64}Zn related to the superallowed Fermi decays of ^{50}Mn and ^{62}Ga have been performed [6–8]. These studies highlight the power of combining different techniques and the complementarity of studies for fundamental symmetries and nuclear structure.

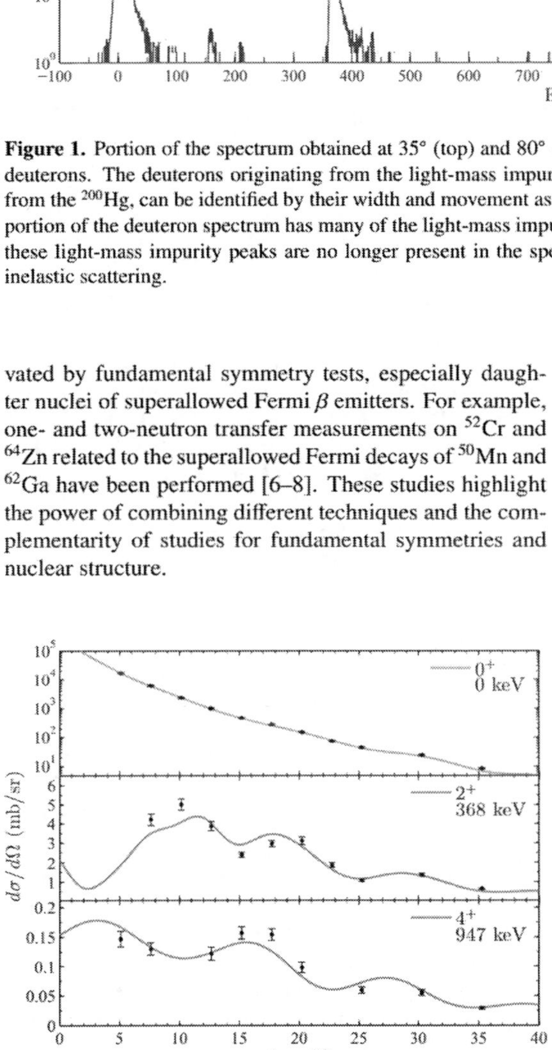

Figure 2. Angular distributions obtained for the ^{200}Hg(d,d') reaction with 22 MeV deuteron beams. The top panel displays the elastic scattering cross sections, the middle panel the inelastic cross section to the 2_1^+ state, and the bottom panel the cross sections for excitation of the 4_1^+ state. The curves are the results of coupled-channel calculations performed with the FRESCO code.

2 Experimental details and results

The experiments to investigate reactions on targets of ^{200}Hg were performed at the Maier-Leibnitz Laboratorium of the Ludwig-Maximilians Universität München and the Technische Universität München. Beams of 22 MeV deuterons up to 1 μA of current bombarded targets of ^{200}Hg ^{32}S of varying thicknesses, but generally on the order of 100 μg/cm^2 of ^{200}Hg. The products of the reaction were momentum analyzed with a Q3D spectrograph employing a position-sensitive focal-plane detector that also provided signals used for ΔE *vs.* ΔE and ΔE *vs.* E for particle identification. The HgS compound was sandwiched between two C foils of 12 μg/cm^2 thickness. While care was taken to minimize the quantity of impurities that the target material was exposed to, as shown in Fig. 1 several impurities beyond the usual appearance of atmospheric gases and ^{28}Si were still observed, albeit for many at the level of a few parts per 10^4 or 10^5. Due to the differences in the kinematics of the deuterons scattering from these light-mass targets *vs.* the heavy ^{200}Hg, the apparent positions of the peaks move in the spectrum to positions corresponding to high Hg excitation energies as the scattering angle increases, and the peaks themselves become increasingly wide as the target mass decreases.

Figure 2 displays the angular distributions obtained for elastic scattering, and excitation of the 2_1^+ and 4_1^+ levels. The smooth curves are results of coupled-channel calculations performed with the FRESCO code [9] using optical model parameters of Daehnick *et al.* [10]. (Different global optical model parameter sets were tested, but the Daehnick *et al.* [10] set provided the best reproduction of the elastic scattering data.) The results of the present

DOI: 10.1051/epjconf/201612303003

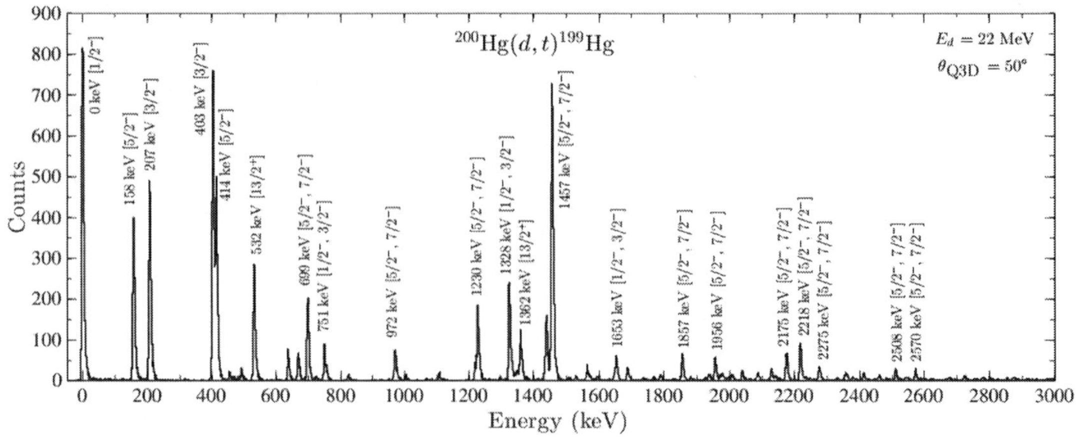

Figure 3. Portion of the triton spectrum observed at an angle of 50° following the ^{200}Hg(d,t) reaction with 22 MeV deuterons. Some of the more prominent peaks are labelled with their corresponding excitation energies in ^{199}Hg and the state J^{π} values.

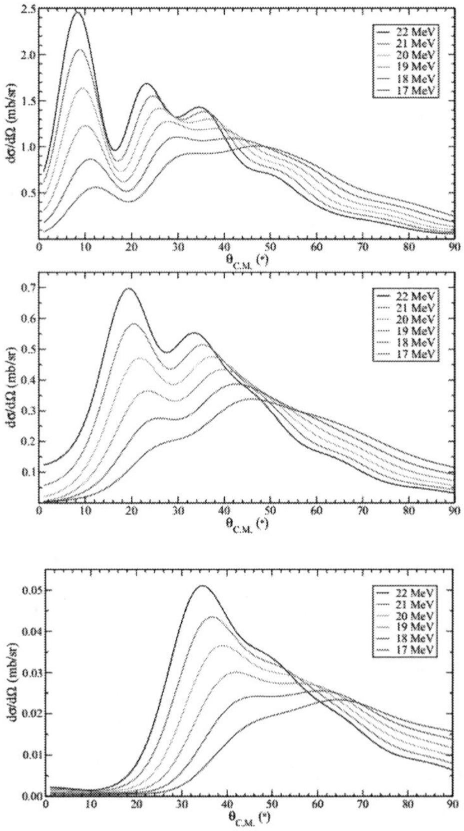

Figure 4. Calculated ^{200}Hg$(d,t)^{199}$Hg angular distributions for the $l = 1$ transition to the $\frac{1}{2}^-$ ground state (top), for $l = 3$ to the 158-keV $\frac{5}{2}^-$ state (middle), and for $l = 6$ to the 532-keV $\frac{13}{2}^+$ state (bottom) for the deuteron beam energies indicated.

work are generally in good agreement with previous results as listed in the Nuclear Data Sheets [11]. The values of $\beta_2 = 0.116(12)$ for excitation of the 2_1^+ level, and $\beta_2 = 0.082(8)$ and $\beta_4 = 0.013(3)$ for excitation of the 4_1^+ level were determined. As Fig. 2 shows, very good fits to the data were obtained. The previously assigned 3^- states at 2151 keV and 2609 keV were observed, and a number of higher-lying states were favoured to be 3^- states as well. The 3_2^- level was the strongest populated, with the 3_1^- level a factor of 7 weaker and the remaining 3^- states weaker still.

Figure 3 shows the spectrum obtained at 50° for the ^{200}Hg(d,t) reaction with 22 MeV deuterons. An excellent resolution of 7 keV full width at half maximum was achieved enabling, for example, the separation of the 403- and 414-keV states. Compared to the previous work by Moyer [12], the angular distributions obtained in the present study have a much enhanced diffraction pattern, allowing the better determination of the transferred l value. Figure 4 displays the calculated angular distributions at incident deuteron beam energies from 17 MeV to 22 MeV for the $J^{\pi} = \frac{1}{2}^-$ ground state, populated by an $l = 1$ transfer, the $J^{\pi} = \frac{5}{2}^-$ 158-keV state, populated by an $l = 3$ transfer, and the $J^{\pi} = \frac{13}{2}^+$ 532-keV state, populated by an $l = 6$ transfer. The experimental angular distributions for these corresponding states are shown in Fig. 5 with the calculated angular distributions. Excellent agreement between the experimental data and the theoretical curves are observed. From the fits to the data, the spectroscopic strengths, defined by

$$S_{jl} = \frac{\left.\frac{d\sigma}{d\Omega}\right|_{\text{exp}}}{\left.\frac{d\sigma}{d\Omega}\right|_{\text{s.p.}}} \qquad (1)$$

are extracted where $\left.\frac{d\sigma}{d\Omega}\right|_{\text{s.p.}}$ is the calculated single-particle cross section using the FRESCO code, and optical model

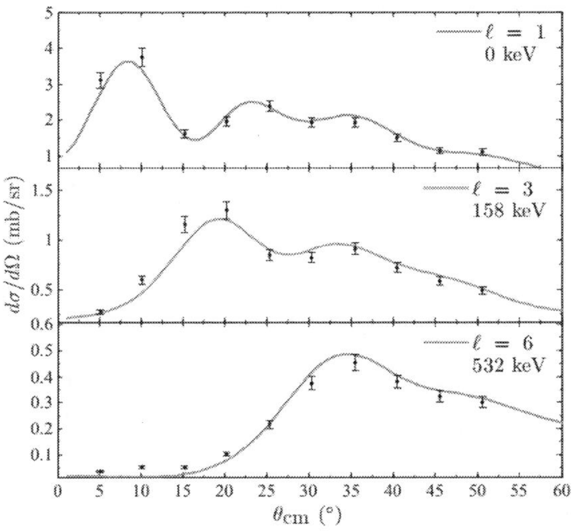

Figure 5. Observed (d, t) angular distributions for population of the ^{199}Hg $\frac{1}{2}^-$ ground state (top), 158-keV $\frac{5}{2}^-$ (middle) and 532-keV $\frac{13}{2}^+$ (bottom). The curves are results of FRESCO calculations for the transferred l value indicated.

parameters from Refs. [10] and [13] for the deuterons and tritons, respectively.

In total, 91 states, of which 50 were newly discovered, in ^{199}Hg were observed up to an excitation energy of approximately 3 MeV. Of these 50 new levels, 22 have definitely assigned transferred l values, with a further 15 previously identified levels also having firm transferred l values. For the remaining levels, their population is so weak that definite assignments could not be made. The spectroscopic strengths extracted for $l = 1$, $l = 3$, and $l = 6$ transfer were combined with previous results from the ^{200}Hg(d, p) reaction by Moyer [12] to test the sum rule. The sum rule for a specific single-particle orbital with angular momentum j is

$$\sum_{\text{pickup}} S_{jl} + \sum_{\text{stripping}} (2j+1) S_{jl} = 2j+1 \qquad (2)$$

if all possible single-particle strength is observed. Since no distinction can be made between the $l = j \pm \frac{1}{2}$ transfers, the combined sum rule for $l = 1$ transfer should be equal to 6, for $l = 3$ it should yield 14, and for $l = 6$ it should equal 14. The experimentally observed sums are 6.4, $\approx$11, and 14, respectively, each with approximately 20% uncertainty. Thus, all the expected strength has been observed, except for a small amount of $l = 3$ strength.

3 Summary

New experiment data on (d, d') inelastic scattering on targets of ^{200}Hg have been obtained and analyzed with the FRESCO coupled-channel code. The new data indicate that the 3^-_2 level at 2609 keV is the strongest populated 3^- state, accounting for at least 50% of the observed 3^- excitation strength. In addition, new results obtained for the ^{200}Hg$(d, t)^{199}$Hg reaction has located nearly all of the $p_{1/2,3/2}$, $f_{5/2,7/2}$, and $i_{13/2}$ single-particle strength. Similar data on targets of ^{198}Hg are currently under analysis, and future work will include inelastic scattering studies on targets of ^{199}Hg and γ-ray spectroscopy to elucidate the $E1$ strength in low-lying levels of the Hg isotopes.

Acknowledgements

This work was supported in part by the Natural Sciences and Engineering Research Council (Canada). TRIUMF receives federal funding via a contribution agreement through the National Research Council of Canada.

References

[1] A. D. Sakharov JETP Lett. **5**, 24 (1967).
[2] M. E. Peskin and D. V. Schroeder, *An Introduction To Quantum Field Theory (Frontiers in Physics)* (Westview Press, 1995), ISBN 0201503972.
[3] J. Pendlebury and E. Hinds, Nucl. Instrum. Meth. **A440**, 471 (2000).
[4] B. Graner, Y. Chen, E. G. Lindahl, and B. R. Heckel, arXiv:1601.04339 [physics.atom-ph].
[5] V. V. Flambaum and V. G. Zelevinsky, Phys. Rev. C **68**, 035502 (2003).
[6] K. G. Leach, P.E. Garrett, G.C. Ball *et al*, Phys. Rev. C **94**, 011304(R) (2016).
[7] K. G. Leach, P. E. Garrett, I. S. Towner, *et al*, Phys. Rev. C **87**, 064306 (2013).
[8] K. G. Leach, P. E. Garrett, C. E. Svensson *et al*, Phys. Rev. C **88**, R031306 (2013).
[9] I. J. Thompson, Comp. Phys. Rep. **7**, 167 (1988).
[10] W. W. Daehnick, J. D. Childs, and Z. Vrcelj, Phys. Rev. C **21**, 2253 (1980).
[11] F. Kondev and S. Lalkovski, Nucl. Data Sheets **108**, 1471 (2007).
[12] R. Moyer, Phys. Rev. C **5**, 1678 (1972).
[13] X. Li, C. Liang, and C. Cai, Nucl. Phys. **A789**, 103 (2007).

EPJ Web of Conferences **123**, 03004 (2016) DOI: 10.1051/epjconf/201612303004
Heavy Ion Accelerator Symposium 2015

Probing cluster structures through sub-barrier transfer reactions

D. C. Rafferty[1,a], M. Dasgupta[1], D. J. Hinde[1], C. Simenel[1], E. C. Simpson[1], E. Williams[1], I. P. Carter[1], K. J. Cook[1], D. H. Luong[1], S. D. McNeil[1], K. Ramachandran[2], K. Vo-Phuoc[1], and A. Wakhle[3]

[1] *Department of Nuclear Physics, Australian National University, Canberra, Australia*
[2] *Nuclear Physics Division, Bhabha Atomic Research Centre, Mumbai, India*
[3] *National Superconducting Cyclotron Laboratory, Michigan State University, East Lansing, MI, USA*

Abstract. Multinucleon transfer probabilities and excitation energy distributions have been measured in ^{16,18}O, ^{19}F + ^{208}Pb at energies between 90% - 100% of the Coulomb barrier. A strong 2p2n enhancement is observed for all reactions, though most spectacularly in the ^{18}O induced reaction. Results are interpreted in terms of the Semiclassical model, which seems to suggest α-cluster transfer in all studied systems. The relation to cluster-states in the projectile is discussed, with the experimental results consistent with previous structure studies. Dissipation of energy in the collisions of ^{18}O is compared between different reaction modes, with cluster transfer associated with dissipation over a large number of internal states. Cluster transfer is shown to be a long range dissipation mechanism, which will inform the development of future models to treat these dynamic processes in reactions.

1 Introduction

The propensity of nucleons within the nuclear medium to coalesce into alpha- particles has been known since the earliest days of nuclear physics. Alpha-decay, in which the Helium nucleus emerges preformed from the parent nuclide was one of the first discoveries in the field. In fact, some of the initial models of nuclear structure (prior to the discovery of the neutron) posited the alpha particle as the basic building block of the nucleus [1]. Such models quickly fell out of favour with the formulation of the liquid drop and shell model of the nucleus, which were better able to predict nuclear properties. However, in recent years cluster models have seen a renaissance, due to their ability to predict certain peculiar spectral properties of some light nuclei.

A classic example of a cluster is the first excited 0^+ state in ^{12}C, known as the Hoyle state [2]; a resonance state of three alpha particles that decays only very rarely to stable ^{12}C, and is the sole production mechanism of this isotope, that is essential to all life on earth. Whilst many studies have examined the properties of cluster resonance states in α-conjugate nuclei (those that are even-even with N = Z), more recently cluster states have been identified in non-conjugate nuclei, with this phenomena seen to be particularly important in neutron-rich nuclei that have become accessible with the advent of radioactive ion beam facilities. The influence of such structures on reaction dynamics remains mostly unexplored.

Coupled channels models have been very successful at reproducing fusion cross sections in the near-barrier region [3], but there are discrepancies both above and far below where cross sections are hindered relative to the predictions [4, 5]. At above barrier energies this has been attributed to dissipative processes that occur during the very violent impact between the ions. Past experiments have shown that dissipation remains important even at sub-barrier energies, so may also have some bearing on the hindrance effects seen in this energy region. Cluster transfer is a possible long-range dissipation mechanism [6], since the cluster states that might be expected to participate are predicted to be radially diffuse [7].

In this work multi-nucleon transfer probabilities and excitation energy distributions in ^{16,18}O, ^{19}F + ^{208}Pb spanning the well-below to above barrier energy regions have been measured. The chosen projectiles have been previously identified to exhibit clustering properties in the context of structure studies [8]. How multi-nucleon transfer varies between the projectiles and how the prominence of different channels evolves as a function of the internuclear separation, as well as how energy is dissipated between the reaction partners has been studied.

2 Experiment

An experiment was performed at the Heavy Ion Accelerator facility at the ANU in June 2013, in which a ^{208}Pb target was bombarded with ^{16}O, ^{18}O, and ^{19}F beams, at energies from 100% to 90% of the Coulomb barrier V_B. The experimental setup is shown in Fig 1. The light projectile-like fragments were detected at 160.6^o, with yields normalized to elastic scattering events in two forward angle monitor detectors to obtain absolute transfer probabilities.

[a] e-mail: dominic.rafferty@anu.edu.au

The detector used in this experiment was a simple Frisch-grid ionization chamber, coupled to a Silicon detector [9]. As ions enter the detector, their energy loss in the gas (ΔE) is recorded together with their residual energy on reaching the Silicon detector ($E_{\rm Si}$). Together these allow a separation of the reaction products in mass and Z.

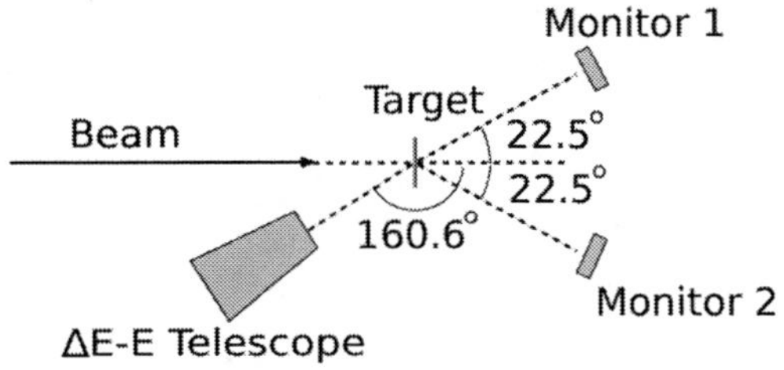

Figure 1. Schematic diagram of experimental setup.

3 Analysis

To identify reaction products, the unique locus of each species must be determined from the $\Delta E - E_{\rm Si}$ spectrum, a typical example of which is shown in Fig. 2(a). The red line in the figure shows the locus of ^{12}C, mapped by tracing the distribution of inelastically scattered ^{12}C from a thick tantalum target. Projecting the $\Delta E - E_{\rm Si}$ spectrum in ΔE allows for a separation of the products within the gated region shown in Fig. 2 by mass. A typical mass separation spectrum is shown in Fig. 2(b), where the products are rebinned according to their relative ΔE compared to a product for which the locus in the $\Delta E - E$ spectrum is well known.

4 Results

The measured transfer probabilities have been interpreted in terms of the Semiclassical model [10]. In this approach, the probability of each mode has the form:

$$P(r_{\rm min}) \propto \exp(-2\alpha r_{\rm min}) \tag{1}$$

Where $\rm r_{min}$ is the distance of closest approach, and α is related to the binding energy of the transfered nucleons. α can be calculated or extracted from transfer probability data, by fitting the exponential slope of the probabilities outside the barrier radius- for the systems studied, this involves fitting the data beyond $r_{min} \geq 13.2$ fm. At the sub-barrier energies studied, it is assumed that the trajectories are purely Coulomb enabling a calculation of $\rm r_{min}$ using:

$$r_{\rm min} = \frac{Z_p Z_t e^2}{4\pi\epsilon}\left(1 + \csc\frac{\theta c.m}{2}\right) \tag{2}$$

where p and t denote projectile and target respectively. In this simple model of independent sequential transfer, the probability for transfer of N nucleons is a simple product of probabilities:

$$P_{Nn} = (P_{1n})^N \tag{3}$$

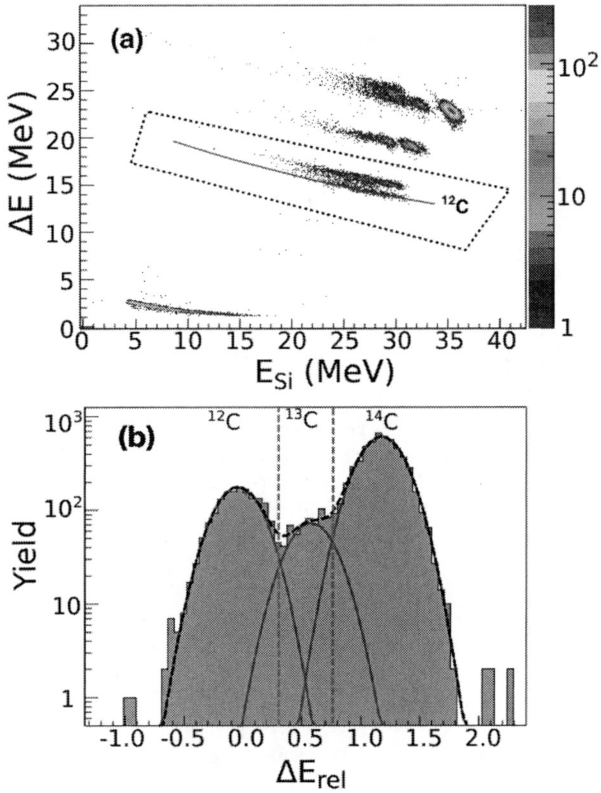

Figure 2. (Colour online) (a) $\Delta E - E_{\rm Si}$ plot obtained in the reaction ^{16}O + ^{208}Pb at $0.98V_B$. The red line shows the ^{12}C locus from which the relative energy loss (ΔE_{rel}) spectrum is calculated. The ΔE_{rel} spectrum is determined from events within the dashed contour. (b) Resulting ΔE_{rel} spectrum. Black dashed curve shows the multiple Gaussian function fitted to the distribution. Red curves indicate the resulting fitted components, which are attributed to yields of the expected isotopes, in this case ^{12,13,14}C. Vertical dashed lines show the gate limits as determined by the intersections between adjacent fitted peaks.

More realistic transfer probabilities can be calculated with advanced microscopic models [11, 12]. In this section transfer probabilities as a function of $\rm r_{min}$ and an example of the extracted excitation energy spectra are shown.

4.1 Transfer Probabilties

4.1.1 $^{16}O + ^{208}Pb$ *: Fig. 3*

Single proton transfer is shown to be dominant over the energy range considered. The semiclassical prediction for 2p transfer is shown as the red dashed line in this figure. This is the square of a fit to the 1p data beyond 13.2 fm, where the loss of flux to fusion does not disturb the exponential slope α. As shown, the two prominent $\Delta Z = 2$ modes shown in the figure are strongly enhanced relative to this prediction. Also interesting is the increasing equiv-

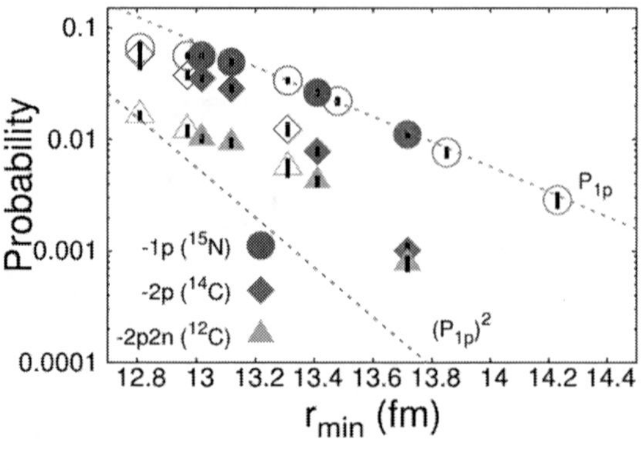

Figure 3. Transfer probabilities in ^{16}O + ^{208}Pb. Full symbols are the measurements reported from the June 2013 experiment. Empty symbols are those measured in a previous experiment [13] with a lower mass resolution. Blue dashed line shows a fit to the 1p data in the range $r_{min} > 13.2$ fm. Red dashed line shows the square of the fitted function, representing the prediction of the Semiclassical model for sequential transfer (see equation 2).

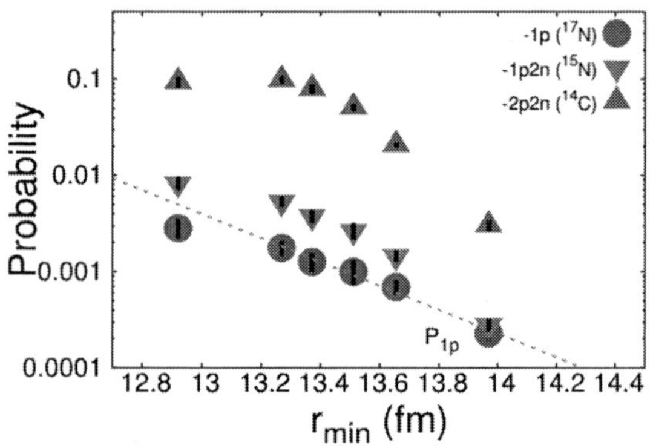

Figure 4. Transfer probabilities in ^{18}O + ^{208}Pb. Blue dashed line shows a fit to the 1p data in the range $r_{min} > 13.2$ fm.

alence of the 2p and 2p2n modes with increasing radial separation.

4.1.2 $^{18}O + ^{208}Pb$: Fig. 4

Single proton transfer is comparatively weak in this reaction. The transfer is dominated by 2p2n transfer over all distances, and also the 1p2n mode is far stronger than 1p, though these become equivalent at larger separations. Again shown is the fit of 1p data, but the semiclassical predictions (e.g. $P_{2p} = (P_{1p})^2$) for all other significant transfer modes in this case are not visible within the range of the figure.

Previous experiments have demonstrated a rich cluster structure in ^{18}O [8, 14, 15], with α resonant states found over a wide excitation range and attributed to a $^{14}C + \alpha$

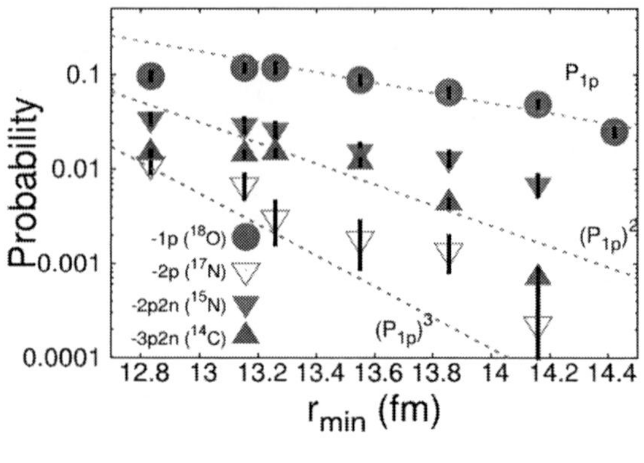

Figure 5. Transfer probabilities in ^{19}F + ^{208}Pb. Blue dashed line shows a fit to the 1p data in the range $r_{min} > 13.2$ fm. Red and green dashed lines are the square and cube respectively of the fitted 1p function, corresponding to the prediction of the semiclassical model (equation 2).

cluster structure. This is fully consistent with such a dominance of 2p2n over other transfer modes.

4.1.3 $^{19}F + ^{208}Pb$: Fig. 5

Single proton transfer is strongest here over the whole energy range. 2p is shown to be very weak, falling in fact lower than the Semiclassical prediction. $\Delta Z = 2$ transfer is dominated by the 2p2n mode. A strong population of events with $\Delta Z = -3$, with a significant yield of ^{14}C at all energies is observed. The 3p2n is most significant, and is equivalent in magnitude to 2p2n at higher energies. This channel falls off quickly however, whilst the 2p2n decays in proportion to 1p with increasing distance.

Previous studies of this reaction at energies closer to the barrier [16] found similar results, and interpreted this behaviour as evidence of clustering in ^{19}F, and in particular that the strong enhancement of 3p2n transfer was indicative of a direct (non-sequential) transfer of p + α. Here it is shown that this enhancement seems to disappear at larger internuclear separations. The strong 2p2n mode, on the other hand, seems to maintain the enhancement down to the deep sub-barrier region.

4.2 Energy Dissipation

The reaction Q-value is reconstructed on an event-by-event basis, from which the excitation energy can be derived. The Q-value can be determined from:

$$Q = \frac{A_3 + A_4}{A_4}E_3 - \frac{A_4 - A_1}{A_4}E_1 - \frac{2\sqrt{A_1 A_3 E_1 E_3}}{A_4}\cos\theta_{lab} \quad (4)$$

Where the subscript notation is in standard reaction form 1(2,3)4, with E denoting energies and A atomic masses. The excitation energy is then obtained from the difference between this value and the ground state Q-value for the reaction mode in question $Q_{g.g}$:

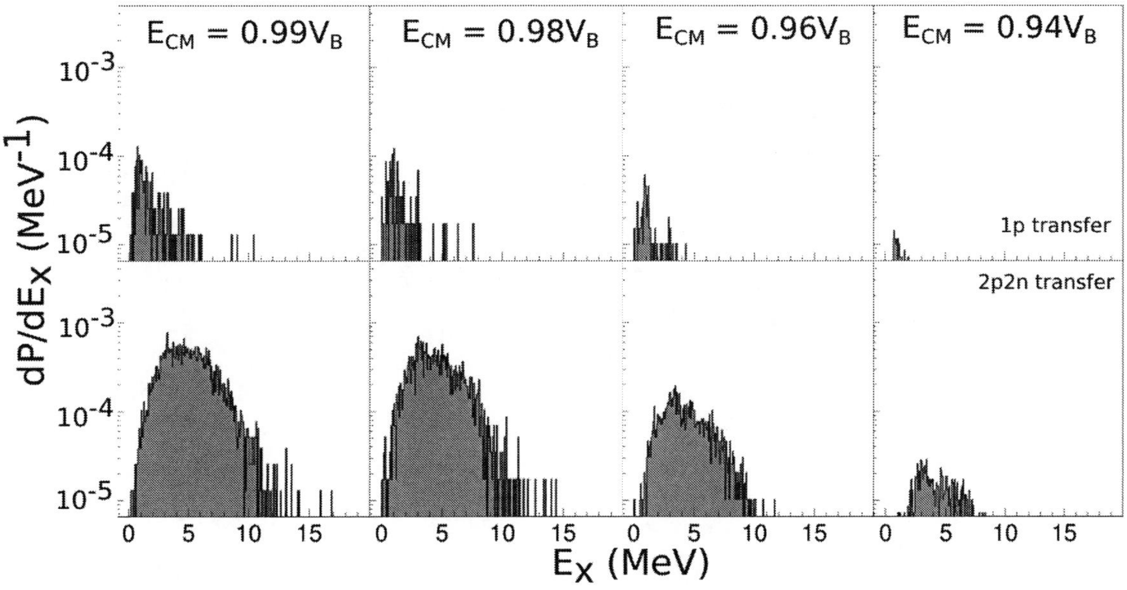

Figure 6. Excitation energy distribution of 1p and 2p2n modes in ^{18}O + ^{208}Pb as it varies with bombarding energy, given in units of the Coulomb barrier energy V_B.

$$E_x = Q_{g.g} - Q \quad (5)$$

In Fig. 6 shown is the recorded E_x spectra for the 1p and 2p2n transfer modes in ^{18}O + ^{208}Pb. Since the target is so much heavier, with a much higher level density than the light projectiles, the target-like fragment is expected to absorb most of the excitation energy in the transfer process. Observed in Fig. 6 is that whilst the transfer of a single proton shows a detailed structure most concentrated at low E_x, the transfer of 2p2n (α) is associated with a strong broad distribution of energy extending up to higher E_x distributions, reaching ~ 10 MeV even at the lowest energy.

5 Conclusions

It has been demonstrated that there are vastly different transfer reactions between the ^{16,18}O, ^{19}F + ^{208}Pb. These effects are likely related to cluster structures that can be found in the projectiles. The very strong 2p2n transfer mode in ^{18}O + ^{208}Pb is consistent with previous investigations into clustering in ^{18}O. ^{16}O, despite being an α-conjugate nucleus, displays comparatively weak 2p2n transfer, though it appears to grow in relation to 2p with internuclear separation. ^{19}F also displays a strong 2p2n mode, that can be attributed to cluster transfer on the basis of the very weak 2p mode.

Excitation energy distributions involving the transfer of multiple nucleons are typically broad and featureless, and it has been shown how in ^{18}O + ^{208}Pb the transfer of an α-particle led to high excitation energies, well above the particle emission thresholds in the target-like fragment, even well below the barrier. While omitted here for brevity, ^{16}O and ^{19}F-induced reactions show similar behaviour. This is a potential mechanism through which energy dissipation can occur at long separations.

It is the goal of this project to establish the systematics of how this differs among light to medium mass projectiles, at energies near to far below the Coulomb barrier. A modified version of the coupled channels code CCFULL [17] is in development to incorporate dissipation into this framework in an attempt to understand the effect of dissipation on fusion cross sections.

Acknowledgements

Thanks go to all of the technical staff at the ANU heavy ion accelerator facility for their support. This work was supported by Australian Research Council grants FT120100760 and FL110100098.

References

[1] L. R. Hafstad and E. Teller, Phys. Rev. **54**, 681 (1938).
[2] F. Hoyle, The Astrophysical Journal Supplement Series **1**, 121 (1954).
[3] K. Hagino and N. Takigawa, Progress of theoretical physics **128**, 1001 (2012).
[4] J. O. Newton, R. D. Butt, M. Dasgupta *et al.*, Phys. Rev. C **70**, 024605 (2004).
[5] M. Dasgupta, D. Hinde, A. Diaz-Torres *et al.*, Physical review letters **99**, 192701 (2007).

[6] A. Diaz-Torres, D. Hinde, M. Dasgupta *et al.*, Physical Review C **78**, 064604 (2008).

[7] Y. Kanada-En'yo, M. Kimura and H. Horiuchi, Comptes rendus Physique **4**, 497 (2003).

[8] E. Johnson, G. Rogachev, V. Goldberg *et al.*, The European Physical Journal A **42**, 135 (2009).

[9] G. F. Knoll, *Radiation detection and measurement* (John Wiley & Sons, 2010)

[10] R. A. Broglia and A. Winther (1991).

[11] C. Simenel, Phys. Rev. Lett. **105**, 192701 (2010).

[12] G. Scamps and D. Lacroix, Phys. Rev. C **87**, 014605 (2013).

[13] M. Evers, M. Dasgupta, D. Hinde *et al.*, Physical Review C **84**, 054614 (2011).

[14] G. Morgan, D. Tilley, G. Mitchell *et al.*, Nuclear Physics A **148**, 480 (1970).

[15] V. Goldberg, K. M. Källman, T. Lönnroth *et al.*, Physics of Atomic Nuclei **68**, 1079 (2005).

[16] D. Biswas, R. Choudhury, B. Nayak *et al.*, Physical Review C **56**, 1926 (1997).

[17] K. Hagino, N. Rowley and A. Kruppa, Computer Physics Communications **123**, 143 (1999).

DOI: 10.1051/epjconf/201612303005

Nuclear structure effects in quasifission – understanding the formation of the heaviest elements

D. J. Hinde[1,a], E. Williams[1], G. Mohanto[1,b], C. Simenel[1], D. Y. Jeung[1], M. Dasgupta[1], E. Prasad[1], A. Wakhle[1,c], K. Vo-Phuoc[1], I. P. Carter[1], K. J. Cook[1], D. H. Luong[1], C. S. Palshetkar[1,d], D. C. Rafferty[1], and E. C. Simpson[1]

[1]*Department of Nuclear Physics, Research School of Physics and Engineering, The Australian National University, Canberra, ACT 2601, Australia*

Abstract. Quasifission is an important process suppressing the fusion of two heavy nuclei in reactions used to create superheavy elements. Quasifission results in rapid separation of the dinuclear system initially formed at contact. Achieving reliable a priori prediction of quasifission probabilities is a very difficult problem. Through measurements with projectiles from C to Ni, the Australian National University's Heavy Ion Accelerator Facility and CUBE spectrometer have been used to map out mass-angle distributions (MAD) - the fission mass-ratio as a function of centre-of-mass angle. These provide information on quasifission dynamics in the least model-dependent way. Average quasifission time-scales have been extracted, and compared with TDHF calculations of the collisions, with good agreement being found. With the baseline information from the survey of experimental MAD, strong influences of the nuclear structure of the projectile and target nuclei can be clearly determined.

1 Introduction

Fusion of two heavy nuclei has been a successful pathway to produce new very heavy and superheavy elements (SHE). However, the production yield of SHE is very significantly suppressed [1] by quasifission [2]. This dynamical non-equilibrium process results when the combined system formed after capture separates before a compact compound nucleus is formed, resulting in two (fission-like) fragments. The probability of quasifission (P_{QF}) can be very large, with the complementary probability of compound nucleus formation ($P_{CN} = 1 - P_{QF}$) being possibly lower than 10^{-6} in unfavourable reactions. Understanding the competition between quasifission and fusion is thus crucial to predict the best fusion reactions to use to form new isotopes of heavy and super-heavy elements. The months of beamtime needed to determine if a given reaction is likely to be successful or not gives additional motivation to obtain a reliable predictive model based on an full understanding of this competition.

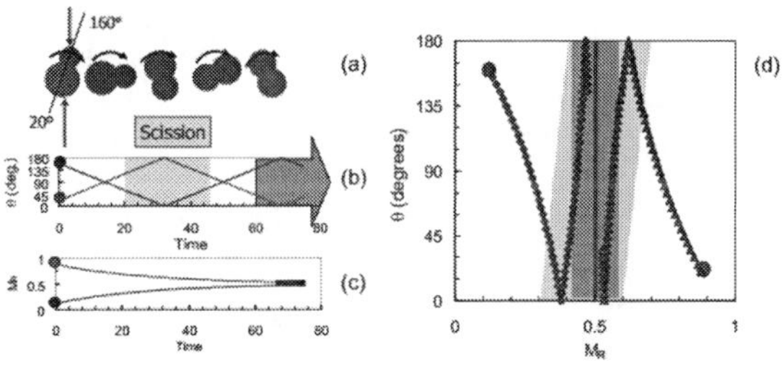

Figure 1. The relationship between sticking time and the mass-angle distribution. For the reaction sequence sketched in (a), the time dependence (arbitrary units) of angle (b) and mass-ratio (c) are shown. Such short times cannot be measured directly, but angle and mass-ratio can. The individual mass and angle dependencies combine to give a trajectory (d) on the MAD for a single impact parameter. Including a range of impact parameters, scission after half a turn (pink) or a full turn (blue) will give strong or weak mass-angle correlations respectively.

2 Mass-Angle Distributions

The measurement of all binary mass-splits (characterised by the mass ratio $M_R = M_1/(M_1 + M_2)$ between the initial projectile mass M_1 and the target mass M_2, when determined over a wide range of scattering angles, results in a two-dimensional matrix, referred to as a mass-angle distribution (MAD). The relationship of the MAD to the "sticking time" between capture and scission is illustrated schematically in Fig. 1. The projectile nucleus (blue) is incident from the top of the page, and sticks to the larger target nucleus (red). The system then rotates, Fig. 1(b) illustrating schematically angle against time (in arbitrary units) for a single angular momentum value; in reality a distribution will be present. For a parabolic potential, mass-symmetry is approached with an expected time dependence $1-\exp(\frac{t}{\tau_{eq}})$, where τ_{eq} is the mass-equilibration time

[a]e-mail: david.hinde@anu.edu.au
[b]Current address: BARC, Mumbai, India
[c]Current address: National Superconducting Cyclotron Laboratory, MSU, East Lansing, Michigan 48824, USA
[d]Current address: TIFR, Mumbai, India

DOI: 10.1051/epjconf/201612303005

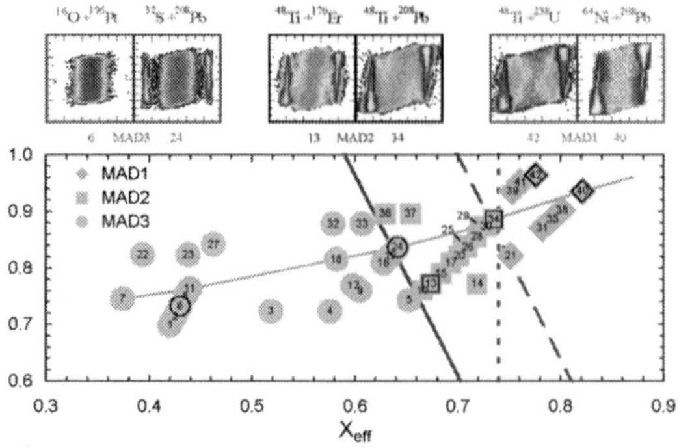

Figure 2. The numbers surrounded by a symbol indicating the classification of MAD observed refer to the reaction number in Ref. [3], and are plotted as a function of the effective fissility parameter in the entrance channel x_{eff} and the compound nucleus fissility parameter x_{CN} (see text). The blue diagonal full line represents the empirical boundary between reactions with no mass-angle correlation (to the left) and those that have (right). The thin line represents the locus of reactions with ^{208}Pb (see text). Examples of MAD are shown in the panels above.

constant [2]. This dependence is sketched in Fig. 1(c). As a dynamical out-of-equilibrium process, the time scale information carried by the MAD gives key information on the dynamics.

It has been found that the quasifission probability and MAD characteristics can depend strongly on the nuclear structure of the two colliding nuclei. This may not be surprising, since nuclear fusion requires the merger of two individual quantum systems, each with their own individual shell structure. In particular, heavy statically deformed nuclei show [4–7] suppression of fusion when the long axis is aligned with the contact angle of the lighter nucleus (called deformation alignment). This condition applies in collisions at sub-barrier energies. In contrast, collisions of spherical magic nuclei at sub-barrier energies can result in enhancement of heavy element yields [8, 9] and fission distributions characteristic of fusion-fission [10, 11].

To improve our quantitative understanding of the role of shell structure in the dynamics of quasifission, we make an analogy with the liquid drop model approach to nuclear masses, in which localized shell effects can be quantified when the underlying smooth (liquid drop) trends are well defined. To define the smooth trends in quasifission, a large number of MAD measurements have been selected from measurements recently made at the ANU [3–5, 11–17]. By investigating empirically those nuclear stucture variables that affect quasifission, the ultimate goal to have a reliable predictive model of quasifission, including all relevant physics, will be a step closer.

During their development, such models must be tested. To test by comparing with measured superheavy element cross sections is appealing. However, comparison with measured characteristics of the competing quasifission process seems more reliable: if a model can't reproduce the quasifission characteristics, which may comprise 99% or more of the cross section, it may throw doubt on the reliability of predictions of the fusion probability, which makes up the remaining 1%, or even less.

To determine the smooth trends in quasifission dynamics, we map the reaction outcomes (defined in terms of the MAD classes discussed below) against variables that reflect the balance between nuclear and Coulomb forces during the collision. These are expected to determine the dynamics in reactions forming very heavy elements. According to the characteristics of the MAD (minimum mass yield at symmetry with mass-angle correlation, mass-angle correlation with peak yield at symmetry, and no significant mass-angle correlation), they are assigned as type MAD1, MAD2 and MAD3 respectively [3]. Note that it has recently been shown [18] that heavy-ion fusion-fission unexpectedly can show a mass-asymmetric (double-peaked) fission mass distribution, so it is the lack of correlation of mass with angle that characterises slow fission, rather than a peak at mass-symmetry. These MAD categories can be associated with characteristic sticking times [3, 16] of $\tau_{\mathrm{MAD1}} < 5 \cdot 10^{-21}$ s, $\tau_{\mathrm{MAD2}} \sim 5 \cdot 10^{-21}$ s and $\tau_{\mathrm{MAD3}} > 10 \cdot 10^{-21}$ s respectively.

It was proposed in Ref. [19] that there should be scaling behavior for reactions, based on the schematic "chaotic regime dynamics" model of fusion of heavy nuclei. This model also predicted the "extra-extra-push" kinetic energy [20] needed to overcome the unconditional saddle-point energy for a given reaction. The two scaling parameters expected to strongly define the reaction outcome are (i) the effective fissility parameter of the entrance channel, x_{eff} (relevant to necked shapes close to the generally mass-asymmetric contact configuration in the entrance channel), and (ii) the fissility parameter of the compound nucleus, x_{CN} (applicable to shapes without a constricted neck [20] where the mass-asymmetry degree of freedom is not constrained). Experimental MAD outcomes are shown in Fig. 2. Importantly these are chosen for beam energies somewhat above the capture barrier (typically by ~ 6%). Here the experimentally observed effects of deformation alignment [4–7] and closed shells, seen in measurements at below-barrier energies [10, 11], are much reduced [5, 21, 22].

The MAD classes are clustered into groups, indicating the appropriateness of the scaling parameters. The boundary across which a mass-angle correlation becomes significant (between MAD classes 3 and 2) shows a dependence on both variables, with a stronger dependence on the entrance-channel than on the compound nucleus fissility. The full blue line is an estimate of this boundary based on current data. Mass-angle distributions for reactions on this line should show similar mass-angle correlations, associated with similar reaction trajectories and timescales. The same should be true for reactions on nearby parallel lines. The dashed and dotted red lines indicate the uncertainty in the boundary of reactions (MAD1) which do not show a peak at mass-symmetry in the angle-integrated distribution.

Such similarities in behaviour between reactions with similar fissilities, but different nuclear structure, is not

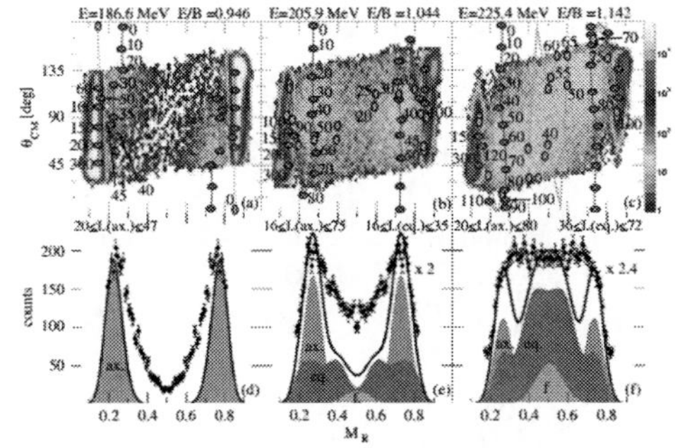

Figure 3. (a)–(c) Measured mass-angle distributions at center-of-mass energies E near the capture barrier for ^{40}Ca+^{238}U. The horizontal and vertical ellipsoids show the final outcome of TDHF calculations for axial and equatorial orientations, respectively. The values of L are indicated near the points associated to the light (heavy) fragments for the axial (equatorial) orientation. (d)–(f) Projected mass ratio M_R in the range $0.2 < M_R < 0.8$. The mass ratio distributions estimated from the TDHF results are shown by shaded areas for quasifission in the axial (ax.) and equatorial (eq.) configurations, and for fusion-fission (f.) events. The sum of these distributions is shown by the solid line for comparison with experiment.

found at sub-barrier energies. Here it is well-known that only collisions with the tips of deformed nuclei lead to capture [23], and that fission angular distributions [6], mass distributions [4, 15, 24] and MAD [5, 17] point to the changing nature (shorter sticking time) of quasifission under these circumstances. Extensive microscopic TDHF calculations of the outgoing masses and angles of binary events have shown a good match [17] to the experimental MAD, and thus to sticking times and mass evolution in the reaction ^{40}Ca+^{238}U. This is illustrated in Fig. 3. This agreement extends throughout the transition from sub-barrier (where contact occurs in the deformation-aligned orientation) to above-barrier energies (at which all collision orientations contribute), as seen in Fig. 3.

In sub-barrier reactions with closed-shell nuclei, it has been shown [11] that increasing the "magicity" in the entrance channel results in narrower mass distributions, correlated with a reduced mass-angle correlation, indicative of longer sticking times. For the ^{48}Ca+^{208}Pb reaction, with four magic numbers, measured mass distributions [10] seem consistent with fusion-fission. These results led to the conclusion [11] that "magicity" plays the strongest role when the N/Z values of the projectile and target nuclei are matched. When this is the case, transfer reactions that destroy entrance-channel magicity (as in the ^{40}Ca+^{208}Pb reaction [11]) are minimized, preserving the closed shell nature of the collision partners as long as possible during the merging of the two nuclei.

Very recent ANU MAD measurements, for the reaction of isotopes of Cr with Pb, are shown in Fig. 4. These measurements were at sub-barrier energies, resulting in the low excitation energies above the ground-state as indicated. They support the picture from the Ca+Pb data.

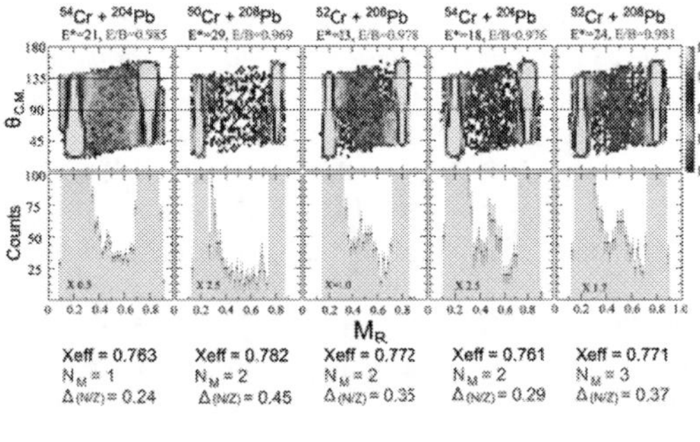

Figure 4. MAD and projected mass-ratio distributions for backward angles from 90° to 135° (as indicated in the MAD plots), for reactions of ^{50,52,54}Cr isotopes with ^{204,206,208}Pb. Sub-barrier beam energies (denoted by E/B), resulted in the low excitation energies E^*. The entrance channel fissility x_{eff}, the number of entrance-channel magic numbers, and the mismatch in N/Z values are indicated for each reaction (see text). The reaction outcome changes from a minimum in yield at mass-symmetry (left) to a narrow peak at symmetry, having no evidence of a mass-angle correlation.

The left three reactions all form the compound nucleus ^{258}Sg. The difference between the N/Z values of the target and projectile nuclei is denoted by $\Delta_{(N/Z)}$ in the figure. The reactions are ordered from less favourable to more favourable for fusion, first by the number of magic numbers in the entrance channel (N_M), and then by $\Delta_{(N/Z)}$. The left-most reaction has only a single magic number in the entrance channel, and shows a U-shaped mass distribution, consistent with MAD class 1. With two magic numbers, the reactions with $\Delta_{N/Z}$ closer to zero show a peak at mass-symmetry, associated with an angle-independent ridge in the MAD. With three magic numbers, but less favourable

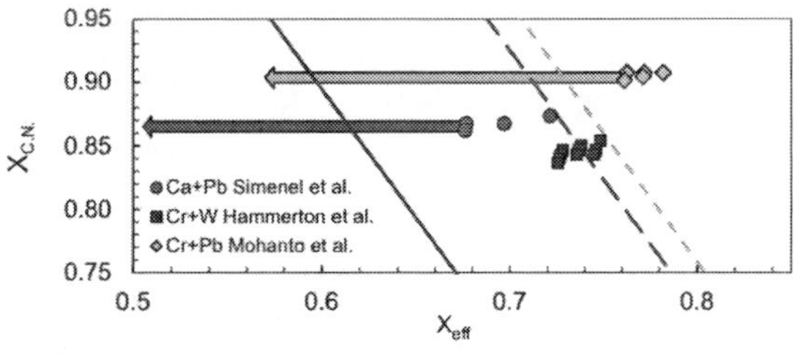

Figure 5. Location of recent measurements of MAD in reactions of isotopes of Ca with Pb, Cr with W and Cr with Pb, plotted as a function of x_{eff} and x_{CN}, as in Fig.2. In the absence of nuclear structure effects, the outcomes should be MAD2 and MAD1. The Cr+W data [25], measured at above-barrier energies, fit well with expectations (but suggest that the MAD1-MAD2 boundary should be shifted slightly to the yellow dashed line). The ^{48}Ca+^{208}Pb reactions show MAD consistent with fusion-fission, as indicated by the arrow. The most surprising outcome is a fraction of Cr+Pb events (where MAD1 is expected) being consistent with MAD3. This suggests that there may be a significant bifurcation of trajectories between fast quasifission, and slow fission which may be consistent in its properties with fusion-fission.

$\Delta_{N/Z}$, a similar result is seen. This mass-symmetric peak becomes a smaller and smaller fraction of the total fission yield (integrated between $0.3 < M_R < 0.7$), as the beam energy is increased. These data show a very similar behaviour to the $^{40,44,48}Ca+^{204,208}Pb$ reactions [11]. However, the transition from a fully U-shaped mass distribution to a U-shaped distribution and a narrow mass-symmetric mass distribution is in some ways an even more drastic change in reaction outcome. It implies a bifurcation of trajectories not seen so clearly in any other measurement of this type. The very sudden change in outcome, and the bifurcation, which depends on magicity and N/Z matching, will be a severe challenge for models of quasifission to reproduce. And yet it is this level of sensitivity that models must strive to reproduce, to optimise experimental opportunities to create new isotopes of superheavy elements.

Acknowledgments

We acknowledge NCRIS operations support for the ANU Heavy Ion Accelerator Facility, and ARC grants DP130101569, DP140101337, FL110100098, FT120100760 and DE140100784. Thanks to N. Lobanov, T. Kibedi and the accelerator staff for supporting use of the superconducting Linac accelerator.

References

[1] D. J. Hinde, M. Dasgupta and A. Mukherjee, Phys. Rev. Lett. **89**, 282701 (2002).
[2] J. Tõke, B. Bock, G. X. Dai *et al.*, Nucl. Phys. A **440**, 327 (1985).
[3] R. du Rietz, E. Williams, D. J. Hinde *et al.*, Phys. Rev. C **88**, 054618 (2013).
[4] D. J. Hinde, R. du Rietz, M. Dasgupta *et al.*, Phys. Rev. Lett. **101**, 092701 (2008).
[5] D. J. Hinde, R. G. Thomas, R. du Rietz *et al.*, Phys. Rev. Lett. **100**, 202701 (2008).
[6] D. J. Hinde, M. Dasgupta, J. R. Leigh *et al.*, Phys. Rev. C **53**, 1290 (1996).
[7] D. J. Hinde, M. Dasgupta, J. R. Leigh *et al.*, Phys. Rev. Lett. **74**, 1295 (1995).
[8] D. J. Hinde and M. Dasgupta, Phys. Lett. B **622**, 23 (2005).
[9] J. Khuyagbaatar, D. J. Hinde, I. P. Carter *et al.*, Phys. Rev. C **91**, 054608 (2015).
[10] E. V. Prokhorova, A. A. Bogachev, M. G. Itkis *et al.*, Nucl. Phys. A **802**, 45 (2008).
[11] C. Simenel, D. J. Hinde, R. du Rietz *et al.*, Phys. Lett. B **710**, 607 (2012).
[12] R. G. Thomas, D. J. Hinde, D. Duniec *et al.*, Phys. Rev. C **77**, 034610 (2008).
[13] R. Rafiei, R. G. Thomas, D. J. Hinde *et al.*, Phys. Rev. C **77**, 024606 (2008).
[14] R. du Rietz, D. J. Hinde, M. Dasgupta *et al.*, Phys. Rev. Lett. **106**, 052701 (2011).
[15] C. J. Lin, R. du Rietz, D. J. Hinde *et al.*, Phys. Rev. C **85**, 014611 (2012).
[16] E. Williams, D. J. Hinde, M. Dasgupta *et al.*, Phys. Rev. C **88**, 034611 (2013).
[17] A. Wakhle, C. Simenel, D. J. Hinde *et al.*, Phys. Rev. Lett. **113**, 182502 (2014).
[18] E. Prasad, D. J. Hinde, K. Ramachandran *et al.*, Phys. Rev. C **91**, 064605 (2015).
[19] J. Blocki, H. Feldmeier and W. Swiatecki, Nucl. Phys. A **459**, 145 (1996).
[20] S. Bjornholm and W. Swiatecki, Nucl. Phys. A **391**, 471 (1982).
[21] K. Nishio, H. Ikezoe, S. Mitsuoka *et al.*, Phys. Rev. C **77**, 064607 (2008).
[22] I. M. Itkis, E. M. Kozulin, M. G. Itkis *et al.*, Phys. Rev. C **83**, 064613 (2011).
[23] M. Dasgupta, D. J. Hinde, N. Rowley *et al.*, Annu. Rev. Nucl. Part. Sci. **48**, 401 (1998).
[24] K. Nishio, H. Ikezoe, I. Nishinaka *et al.*, Phys. Rev. C **82**, 044604 (2010).
[25] K. Hammerton, Z. Kohley, D. J. Hinde *et al.*, Phys. Rev. C **91**, 041602 (2015).

EPJ Web of Conferences **123**, 03006 (2016) DOI: 10.1051/epjconf/201612303006
Heavy Ion Accelerator Symposium 2015

Mass-asymmetric fission in the ^{40}Ca+^{142}Nd reaction

E. Prasad[1,a,b], D. J. Hinde[1], E. Williams[1], M. Dasgupta[1], I. P. Carter[1], K. J. Cook[1], D. Y. Jeung[1], D. H. Luong[1], S. McNeil[1], C. S. Palshetkar[1], D. C. Rafferty[1], C. Simenel[1], A. Wakhle[1,c], K. Ramachandran[1,d], J. Khuyagbaatar[2,e], Ch. E. Dullmann[2,f], B. Lommel[2], and B. Kindler[2]

[1]*Department of Nuclear Physics, Research School of Physics and Engineering, Australian National University, Canberra, ACT 2601, Australia.*
[2]*GSI Helmholtzzentrum fur Schwerionenforschung, 64291, Darmstadt, Germany.*

Abstract. Shell effects play a major role in fission. Mass-asymmetric fission observed in the spontaneous and low energy fission of actinide nuclei was explained by incorporating the fragment shell properties in liquid drop model. Asymmetric fission has also been observed in the low energy fission of neutron-deficient ^{180}Hg nuclei in recent β-delayed fission experiments. This low-energy β-delayed fission has been explained in terms of strong shell effects in pre-scission configurations associated with the system after capture. Calculations predicted asymmetric fission for heavier Hg isotopes as well, at compound nuclear excitation energy as high as 40 MeV. To explore the evolution of fission fragment mass distribution as a function of neutron and proton numbers and also with excitation energy, fission fragment mass distributions have been measured for the ^{40}Ca+^{142}Nd reaction forming the compound nucleus ^{182}Hg at energies around the capture barrier, using the Heavy Ion Accelerator Facility and CUBE spectrometer at the Australian National University. Mass-asymmetric fission is observed in this reaction at an excitation energy of 33.6 MeV. The results are consistent with the β-delayed fission measurements and indicate the presence of shell effects even at higher exciation energies.

1 Introduction

Nuclear fission is a dynamic process involving large scale collective motion, often affected by a subtle interplay of macroscopic and microscopic effects during the transition of the fissioning nucleus from its ground state deformation to the scission point. Fragment mass distributions are an important observable in fission which provides crucial information about the potential-energy landscape of the fissioning system. Fragment mass distributions have been observed to be predominantly asymmetric in the spontaneous or low energy fission of most of the actinide nuclei [1], which could not be explained solely by using the liquid drop model (LDM) [2]. The experimental observations were explained by incorporating the fragment shell properties near the scission configuration, particularly the spherical shell closure (Z = 50, N = 82) or deformed neutron shells (N = 88) [1, 3], within the LDM frame work.

Recent observations of mass-asymmertic fission of the Hg isotopes in β-delayed fission experiments [4, 14] indicated the role of shell structures other than fragment shells in deciding the fission outcomes. The low yield of symmetric fission in ^{180}Hg populating ^{90}Zr (Z = 40, N = 50) led to the speculation that fragment shell effects, though significant in the potential energy surface near scission, must not play a major role in determining the mass split in this nucleus.

Even though β-delayed fission experiments allowed us to access the low energy fission regime of the neutron-deficient nuclei in the mass 180-200 region, the availability of nuclei undergoing this process is severely limited by the stringent conditions on β-decay Q-values and branching ratios [4]. Also, the maximum available excitation energy for the fissioning system is restricted by the Q-value of the parent nucleus [4]. In the actinide region, it is observed that [1, 5] the fragment mass division becomes symmetric with increasing excitation energy, as shell effects generally wash out at higher excitations. In order to understand the evolution of fission fragment mass distribution as a function of neutron and proton numbers (N and Z) of the fissioning system and the excitation energy in the neutron deficient Hg region, we measured the fission fragment mass distributions of ^{182}Hg populated through the fusion reaction ^{40}Ca+^{142}Nd at energies around the capture barrier. The results are compared with the latest experimental data available in the same mass region for the fis-

[a]e-mail: prasad.edayillam@anu.edu.au
[b]Permanent Address: Department of Physics, Central University of Kerala, Kasaragod, 671314, India.
[c]Presently at National Superconducting Cyclotron Laboratory, Michigan State University, Michigan, 48824, USA.
[d]Presently at Bhabha Atomic Research Center, Mumbai 400085, India.
[e]Helmholtz Institute Mainz, 55099 Mainz, Germany.
[f]Institute for Nuclear Chemistry, Johannes Gutenberg University Mainz, 55128 Mainz, Germany.

DOI: 10.1051/epjconf/201612303006

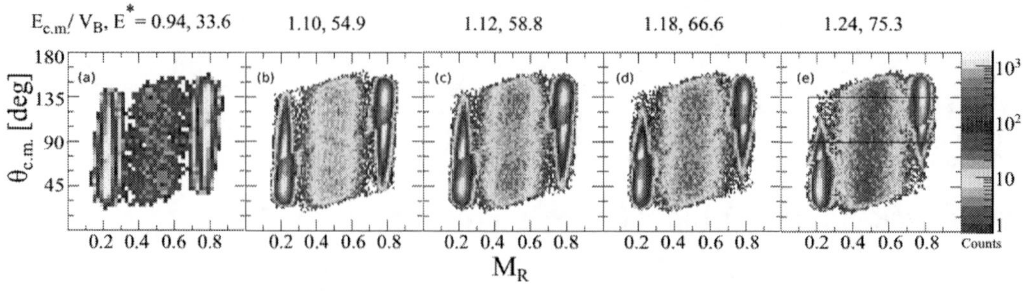

Figure 1. Experimental MAD scatter plots for the $^{40}Ca+^{142}Nd$ reaction at different beam energies. The $\frac{E_{c.m.}}{V_B}$ values and CN excitation energies E^* are given at the top of each plot. E^*, $E_{c.m.}$ and V_B are in units of MeV.

sion of different Hg isotopes and also with the predictions of some of the theoretical calculations available.

2 Experimental details and data analysis

The experiment was performed using the 14UD Pelletron accelerator and superconducting linear accelerator facility at the Australian National University. Pulsed ^{40}Ca beams with a pulse seperation of 106.7 ns and full width at half-maximum (FWHM) of 0.7-1.5 ns were used to bombard the isotopically enriched $^{142}NdF_3$ target of thickness 400 $\mu g/cm^2$, with a 18 $\mu g/cm^2$ carbon backing facing downstream. The measurements were performed at 167.7, 194.9, 199.9, 210.0 and 221.1 MeV laboratory energies and the fission fragments were detected using the CUBE [6] fission spectrometer. CUBE consists of two large area position sensitive multiwire proportional counters, each with an active area of 279 × 357 mm^2. These detectors were mounted at 45° and 90° scattering angles with respect to the beam direction, such that the normal to the central foil of both detectors was at 180 mm from the center of the target. The target normal was oriented at 60° with respect to the beam axis to avoid shadowing of the detectors. Two silicon monitor detectors were mounted at $\theta = 30°$, at azimuthal angles of 90° and 270°, to detect the elastically scattered beam particles.

The CUBE detectors provide timing, position, and energy loss signals of the particles entering them. The calibrated position and time of flight information were used to obtain the fragment velocities and center-of-mass angles. The mass ratio (M_R) of the fragments is then obtained as,

$$M_R = \frac{m_1}{m_1 + m_2} = \frac{v_2}{v_1 + v_2} \quad (1)$$

where m_1 and m_2 are the masses of fragments at scission and v_1 and v_2 are their center-of-mass velocities, respectively. Light particle evaporation from the fragments does not alter the centroids of the M_R distribution, as the mass ratio is derived from the fragment velocities, which - on average - are not affected by the evaporation of light particles.

Mass-angle distribution (MAD) plots [7] of the fragments provide a direct indication of the presence of processes other than fusion-fission in heavy ion reactions. MAD plots generally consist of groups originating from elastic, inelastic, fusion-fission and quasifission processes. Unlike the elastic and inelastic events, quasifission events overlap significantly with those from the fusion-fission making their separation impossible in most cases. However, in general, a strong component of quasifission may be inferred from the presence of mass-angle correlations in a MAD plot [7].

3 Results

MADs of the fragments were generated to probe the presence of processes other than fusion-fission in the $^{40}Ca+^{142}Nd$ reaction and are shown in Fig. 1. The $\frac{E_{c.m.}}{V_B}$ values and compound nucleus (CN) excitation energies (E^*) are given at the top of each plot, where $E_{c.m.}$ is the center-of-mass energy and V_B is the capture barrier from a calculation [8] optimised for heavy systems. Since we detect both the fragments in our experimental setup, the MAD is populated twice, at (M_R, $\theta_{c.m.}$) and ($1 - M_R$, π-$\theta_{c.m.}$). Hence the MAD plots shown in Fig. 1 are suitably mirrored. The non-appearance of a strong correlation in fragment mass and emission angle in the MAD for the $^{40}Ca+^{142}Nd$ reaction indicates the absence of fast quasifission originating from quick re-separation of the di-nuclear system after capture in this reaction. The M_R

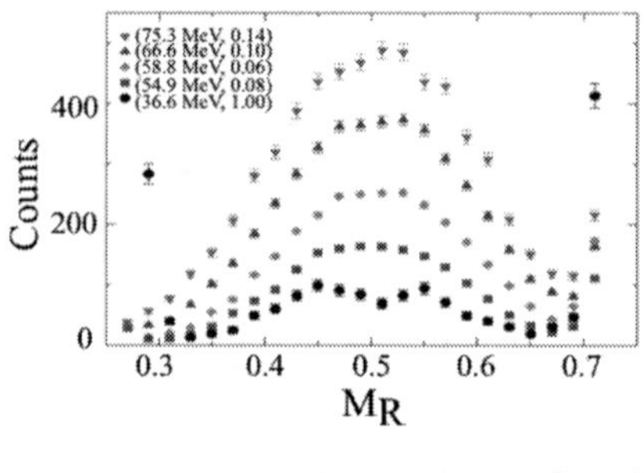

Figure 2. The fission M_R distributions from the $^{40}Ca+^{142}Nd$ reaction at different CN excitation energies as indicated. The yields are scaled to show all energies in a single plot, scaling factors are shown inside the parantheses.

DOI: 10.1051/epjconf/201612303006

distributions of the fragments were obtained by projecting these MADs onto the M_R axis. An angular window ($\Delta\theta$) of 45° [90° $\leq \Delta\theta \leq$ 135° as indicated in Fig. 1 (e)] has been selected to avoid the biasing of the data due to the geometrical limits of the detector and also to eliminate any mirrored data in the final M_R distribution.

The experimental M_R distributions for the ^{40}Ca+^{142}Nd reaction at different CN excitation energies (E^*) are shown in Fig. 2. A strong mass-asymmetric component is observed in the M_R distribution at E^*=33.6 MeV. The M_R distributions at higher excitation energies do not show any direct signature of mass-asymmetric fission and can be represented by a single flat-topped Gaussian function.

The width of the fragment M_R distribution is obtained by fitting the distribution with a single Gaussian function. The standard deviations (σ_{MR}) of the Gaussian fit at different CN excitation energies are shown in Table 1. The σ_{MR} values increase with increasing excitation energy. Even though the absence of mass-angle correlation rules out fast quasifission in the ^{40}Ca+^{142}Nd reaction, quasifission with longer sticking times may be contributing to larger σ_{MR} values ($\sigma_{MR} \geq 0.07$ [7]) observed in this reaction. Slow quasifission could contribute to an increase in mass widths at higher energies and angular momenta, but the smallest contribution is expected at the lowest energy. Because of their longer sticking times in comparison with the fast quasifission events [7], slow quasifission is not expected to produce mass-angle correlations in the MAD plots.

Table 1. Single Gaussian widths (σ_{MR}) obtained for the M_R distributions for the ^{40}Ca+^{142}Nd reaction at different CN excitation energies.

E* (MeV)	σ_{MR}
33.6	0.0862±0.0030
54.9	0.0824±0.0007
58.8	0.0877±0.0006
66.6	0.0950±0.0008
75.3	0.1006±0.0010

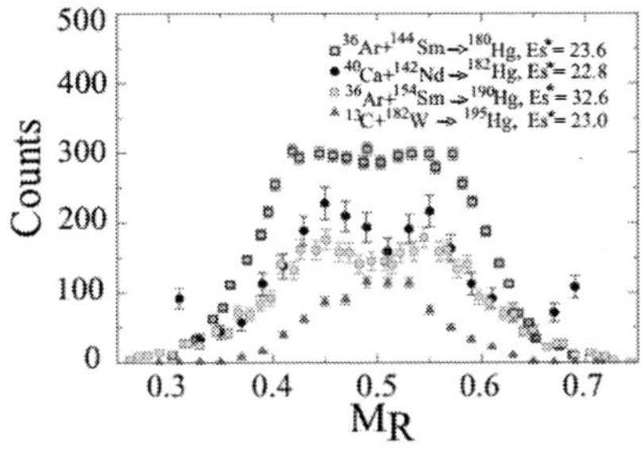

Figure 3. The scaled M_R distributions of fission from different reaction populating isotopes of Hg. The excitation energy available above the zero angular momentum fission barrier (E_s^* in MeV) is also shown.

4 Discussion

According to a recent macroscopic-microscopic finite-range liquid-drop model calculation [9], the zero angular momentum fission barrier height of ^{182}Hg is 10.85 MeV. Hence, the available excitation energy above the saddle point (E_s^*) for ^{182}Hg nuclei formed in the ^{40}Ca+^{142}Nd reaction is 22.8 MeV at the lowest measured energy. The M_R distributions from the fission of ^{180}Hg, ^{190}Hg and ^{195}Hg populated at similar or nearby excitation energies above the zero angular momentum saddle point, populated through the reactions ^{36}Ar+^{144,154}Sm [10] and ^{13}C+^{182}W [11, 12], respectively, are compared with our present observation in Fig. 3. Different parameters of the reactions studied in this paper are given in Table 2. The fragment mass distributions reported in Ref. [10] for the ^{36}Ar+^{144,154}Sm reaction have been converted into M_R distributions in this plot for a direct comparison. It is interesting to note that the fission from ^{180}Hg and ^{190}Hg nuclei produced in the ^{36}Ar+^{144,154}Sm reactions also show signatures of asymmetric mass split. However, a symmetric split has been observed in the fission of ^{195}Hg formed in the ^{13}C+^{182}W reaction, showing a transition from asymmetric to symmetric mass division with increase in neutron number in Hg isotopes. Similarly, a symmetric-peaked distribution has also been reported for the fission of ^{198}Hg [13].

Table 2. The CN produced, CN excitation energy (E^*) and excitation energy available above the zero-angular momentum saddle point (E_s^*) for different reactions studied in this paper.

Reaction	CN	E^* (MeV)	E_s^* (MeV)
^{36}Ar+^{144}Sm [10]	^{180}Hg	33.4	23.6
		39.8	30.0
		47.8	38.0
		65.5	55.7
^{40}Ca+^{142}Nd	^{182}Hg	33.6	22.8
		54.9	44.1
		58.8	48.0
		66.6	55.8
		75.3	64.5
^{36}Ar+^{154}Sm [10]	^{190}Hg	47.8	32.6
		55.9	40.7
		62.4	47.2
		70.5	55.3
^{13}C+^{182}W	^{195}Hg	41.8	23.0
		44.6	25.8
		47.4	28.6

The observation of asymmetric mass division in the low energy fission of ^{180}Hg [4, 14] in β-delayed fission experiments triggered special interest in the fission of

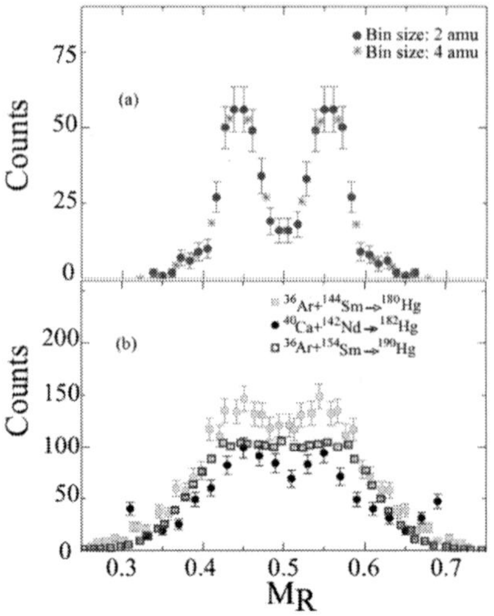

Figure 4. (a) The M_R distributions of ^{180}Hg from the β-delayed fission measurements (E^* = 10.44 MeV). The fragment mass distribution is converted into M_R distribution, with different bin sizes. (b) The M_R distributions of fission from ^{180}Hg (^{36}Ar+^{144}Sm, E^*= 33.4 MeV), ^{182}Hg (^{40}Ca+^{142}Nd, E^*=33.6 MeV) and ^{190}Hg (^{36}Ar+^{154}Sm, E^*= 47.8 MeV).

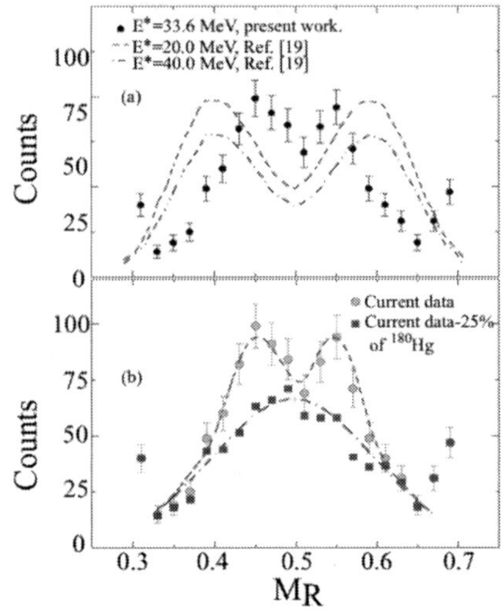

Figure 5. (a) The M_R distributions of fission fragments from the ^{40}Ca+^{142}Nd reaction at E^*=33.6 MeV compared with BSM predictions of ^{182}Hg [19] at E^*=20 and 40 MeV. (b) M_R distribution obtained by subtracting 25% contribution from the fission of ^{180}Hg. A Gaussian fit to this distribution is shown using the dot-dashed line (blue). Dashed line (red) represents the sum of the Gaussian fit values and 25% fission contribution from ^{180}Hg.

nuclei in the neutron deficient Hg region. Soon after this observation, different theoretical explanations were put forward to explain the experimental results. Among these calculations, potential energy surface calculations [4, 15] provided a qualitative explanation of the data and predicted similar observations for heavier isotopes of Hg. Microscopic calculations based on the Skyrme-HFB model [16] also reproduced the experimental observations in ^{180,198}Hg. Calculations based on a modified scission point model [17, 18] also explained low energy fission results observed for ^{180}Hg. Interestingly, though based on different model assumptions, all these calculations reproduced the β delayed fission results reasonably well.

The M_R distributions observed in the ^{40}Ca+^{142}Nd and ^{36}Ar+^{144,154}Sm reactions are thus compared with the results from β-delayed fission in Fig. 4. A striking feature is the similarity in mass division in all these reactions. It is reported that the asymmetric peaks in the fission of ^{180}Hg in Ref. [14] are centered around ^{80}Kr and ^{100}Ru, translating to M_R values 0.46 and 0.54 for the light and heavy fragments, respectively. The M_R peaks observed in the ^{40}Ca+^{142}Nd reaction are also consistent with these values. The peaks of the two Gaussian functions used to represent the mass distributions in the ^{36}Ar+^{144,154}Sm reactions in Ref. [10] also correspond to similar M_R values, consistent with that of β-delayed fission results.

The observation of mass-asymmetric fission in the ^{40}Ca+^{142}Nd fusion-fission reaction appears to support the influence of shell effects up to this excitation energy. The relatively low fissility of ^{182}Hg suggests that the observed fission may be dominated by first chance fission of the CN itself. Fission yield calculations based on Brownian Shape Motion (BSM) by Moller [19] predicted asymmetric fission in Hg isotopes (^{176}Hg to ^{188}Hg) up to CN excitation energies as high as 40 MeV. Similarly, calculations based on finite-temperature density functional theory [20] also predicted the occurrence of mass-asymmetric fission in neutron deficient Hg isotopes at high CN excitation energies up to 30 MeV. We thus compared our results with the BSM predictions for the fission of ^{182}Hg in Fig. 5(a). Though both theoretical predictions and experimental results show similar trends in mass split, the mass centroids of the asymmetric peaks are different. Similar observations may also be made for the low energy [14] and high energy [10] fission of ^{180}Hg. However, the predicted results match with the experimental results for ^{190}Hg [10]. In addition to the mismatch in mass centroids, significant difference in the FWHM of the asymmetric peaks was observed in the experimental results and BSM predictions.

On the other hand, pre-scission neutron multiplicity (ν_{pre}) estimates based on the systematics available in literature (ν_{pre}=0.51 according to the systematics from Ref. [21] and 0.82 from [22]) indicate that multichance fission may be significant in the case of ^{182}Hg, leading to a possibility that low energy fission of ^{180}Hg may be contributing to the observed M_R distribution in the ^{40}Ca+^{142}Nd reaction. Such a possibility is explored by first normalising the area under the M_R distributions from the β-delayed fission study and ^{40}Ca+^{142}Nd reaction at E^*=33.6 MeV and, then, subtracting a fraction of the M_R distribution of ^{180}Hg from that of ^{182}Hg. It was observed that subtracting a 25% contribution of fission of ^{180}Hg from that measured from the ^{40}Ca+^{142}Nd reaction leads to a distribution which can be represented by a single Gaussian centered at mass symmetry. This is shown in Fig. 5(b). While the dot-dashed curve represents the Gaussian fit to the symmetric distribution, the dashed line represents the sum of this Gaussian fit and 25% fraction of asymmetric fission from ^{180}Hg.

In a further step we used the empirical model GEF [23] to compare the experimental results observed in this work. The GEF model [23] uses a large body of experimental information, to develope an empirical, global description of fission quantities. In GEF, the quantitative predictions of fission fragment yields rely on the properties of the fragment shells that define quantum oscillators in mass asymmetry and charge polarisation on the fission path. The positions and shapes of the fission valleys of standard fission channels [24] were obtained by fitting the structures in the measured mass distributions. The potential at elongation is calculated as the sum of the macroscopic potential and the shell effects. Although GEF is not intended for nuclei lighter than Pb, the experimental M_R distribution observed for the ^{40}Ca+^{142}Nd reaction at E_s^*=22.8 MeV has been compared with the GEF predictions for the fission of ^{182}Hg in Fig. 6 (a). The model predictions are in good agreement with the experimental results in terms of the mass centroids, peak to valley ratio and the FWHM of the mass-asymmetric peaks. The model predictions at different E_s^* values for the fission of ^{182}Hg are shown in Fig. 6 (b). It is quite clear that the distribution becomes a single Gaussian-like form at higher excitation energy for ^{182}Hg, resulting from the weakening of shell effects at such high excitations. The experimental M_R distributions obtained for the fission of ^{195}Hg in the ^{13}C+^{182}W reaction at E_s^*= 23.0 and 28.6 MeV are compared with the GEF model predictions in Fig. 6 (c). The model results matches well with the experimental M_R distribution predicting a much narrower M_R distribution for the fission of ^{195}Hg compared with ^{182}Hg. However, GEF also predicts a slightly asymmetric mass-split for the fission of ^{195}Hg, which could not be explored in detail at the moment due to the limited statistics available.

Good agreement between the experimental data and GEF indicates the persistence of shell effects at excitation energies at least up to 34 MeV, consistent with the predictions of Refs. [19, 20]. The success of the GEF model to reproduce the experimental observables suggests that the asymmetric fission observed in the neutron deficient Hg region may also be explained using the same physics that is applicable in the fission of actinide nuclei. This needs further investigation.

From the present results for the ^{40}Ca+^{142}Nd reaction, it is difficult to make definitive statements about the nature of shell effects at excitation energies above E^*=33.6 MeV. Unlike the observation at E^*=33.6 MeV, the signature

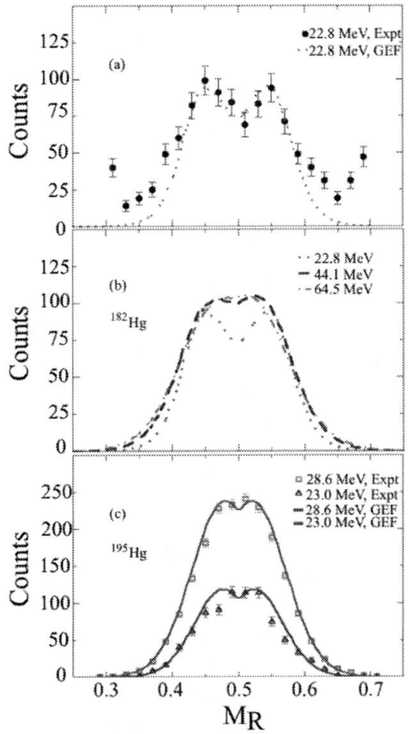

Figure 6. (a) The M_R distributions of fission fragments from the ^{40}Ca+^{142}Nd reaction at E_s^*=22.8 MeV is compared with GEF model predictions. (b) GEF predictions of the M_R distributions for the fission of ^{182}Hg at different E_s^* values. (c) GEF predictions of the M_R distributions for the fission of ^{195}Hg at different E_s^* values.

of mass-asymmetric fission is much weaker at higher excitation energies. The flat-topped M_R distribution observed at E^*=54.9 MeV and above may be indicating some asymmetric component underlying the dominant symmetric distribution. Such a flat-topped distribution has been represented by the sum of two Gaussian functions in Ref. [10] for the case of ^{36}Ar+^{144,154}Sm reactions at E^* up to 65.5 and 70.5 MeV, respectively, for the fission of ^{180}Hg and ^{190}Hg nuclei. More experimental data are required at finer energy steps to conclusively establish the excitation energy dependence of mass division in the fission of neutron-deficient Hg nuclei.

5 Summary and conclusions

In this paper we discussed the fission fragment mass-angle and mass ratio distributions for the ^{40}Ca+^{142}Nd reaction at energies around the capture barrier. Fragment mass-angle and mass ratio distributions were extracted using two-body kinematics [6]. The absence of a strong mass-angle correlations in the MAD clearly rules out the influence of fast quasifission in this reaction. The fission fragment M_R distribution shows an asymmetric mass division with peaks centered around M_R values 0.45 and 0.55 at E^*=33.6 MeV. These M_R values are consistent with the asymmetric mass split observed in the low energy fission of ^{180}Hg in β-delayed fission experiments. Flat-topped M_R distributions have been observed at higher exciation energies, possibly indicating a small component of low E^* asymmetric fission underlying a symmetric component.

The present results are compared with the fission fragment mass distributions from the ^{36}Ar+^{144,154}Sm and ^{13}C+^{182}W reactions populating ^{180,190,195}Hg nuclei, respectively. While the fragments from ^{180,190}Hg showed signatures of asymmetric mass split, a symmetric distrbution is observed for the fission of ^{195}Hg. While the observation of mass-asymmetric fission superficially support the existence of shell effects even at high excitation energies as claimed in [10] and predicted in Refs. [19, 20], we also show empirically that the results could be explained by using a 25% fission contribution from ^{180}Hg. Even though the experimental results show similar trends as predicted in Ref. [19] in mass division, differences have been observed in the position of the mass centroids and FWHM of the peaks. On the otherhand, calculations based on an empirical model [23] reproduced the results perfectly well, supporting the influence of shell effects at compound nuclear excitation energy at least up to 34 MeV. The success of this empirical model also hints that a global description of fission may possibly be able to represent the phenomenon of fission across the nuclear chart.

Acknowledgments

The authors acknowledge support from the Australian Research Council through grants FL110100098, FT120100760, DP130101569, DP140101337, DP160101254 and DE140100784. Support for accelerator operations through the NCRIS program is acknowledged.

References

[1] R. Vandenbosch and J. R. Huizenga, *Nuclear Fission* (Academic, New York, 1973.)
[2] N. Bohr and J. A. Wheeler, Phys. Rev **56**, 426 (1939).
[3] L. Meitner, Nature **165**, 561 (1950).
[4] A. N. Andreyev, M. Huyse, and P. VanDuppen, Rev. Mod. Phys. **85**, 1541 (2013).
[5] I. F. Croall *et al.*, Nucl. Phys. **A 125**, 402 (1969).
[6] D. J. Hinde *et al.*, Phys. Rev. C **53**, 1290 (1996).
[7] R. du Rietz *et al.*, Phys. Rev. C **88**, 034611 (2013).
[8] W. J. Swiatecki *et al.*, Phys. Rev. C **71**, 014602 (2005).
[9] P. Moller *et al.*, Phys. Rev. C **79**, 064304 (2009).
[10] K. Nishio *et al.*, Phys. Lett. B, **748**, 89 (2015).
[11] E. Prasad *et al.*, Phys. Rev. C **91**, 064605 (2015).
[12] K. Ramachandran *et al.*, EPJ web of Conferences **63**, 02017 (2013).
[13] M. G. Itkis, V. N. Okolovich and G. N. Smirenkin, Nucl. Phys. **A 502**, 243c (1989).
[14] J. Elseviers *et al.*, Phys. Rev. C **88**, 044321 (2013).
[15] T. Ichikawa, A. Iwamoto and P. Moller, Phys. Rev. C **85**, 024306 (2012).
[16] M. Warda *et al.*, Phys. Rev. C **86**, 024601 (2012).
[17] A. V. Andreev *et al.*, Phys. Rev. C **88**, 047604 (2013).
[18] S. Panebianco *et al.*, Phys. Rev. C **86**, 064601 (2012).
[19] P. Moller, J. Randrup and A. J. Sierk, Phys. Rev. C **85**, 024306 (2012).
[20] J. D. McDonnel *et al.*, Phys. Rev. C **90**, 021302(R) (2014).
[21] H. Baba *et al.*, J. Phys. Soc. Jpn, **66**, 998 (1997).
[22] M. G. Itkis and A. Ya. Rusanov, Phys. Part. Nucl. **29**, 160 (1998).
[23] K. -H. Schmidt *et al.*, Nuclear Data Sheets **131**, 107 (2016).
[24] U. Brosa *et al.*, Phys. Rep. **197**, 167 (1990).

EPJ Web of Conferences **123**, 04001 (2016)
Heavy Ion Accelerator Symposium 2015

DOI: 10.1051/epjconf/201612304001

Nuclear Forensics using Gamma-ray Spectroscopy

E. B. Norman[1,a]

[1] *Department of Nuclear Engineering, University of California, Berkeley, California 94720, USA*

Abstract. Much of George Dracoulis's research career was devoted to utilising gamma-ray spectroscopy in fundamental studies in nuclear physics. This same technology is useful in a wide range of applications in the area of nuclear forensics. Over the last several years, our research group has made use of both high- and low-resolution gamma-ray spectrometers to: identify the first sample of plutonium large enough to be weighed; determine the yield of the Trinity nuclear explosion; measure fission fragment yields as a function of target nucleus and neutron energy; and observe fallout in the U. S. from the Fukushima nuclear reactor accident.

1 Seaborg's Plutonium?

In early 1941, in Room 307 Gilman Hall on the University of California's Berkeley campus, Glenn Seaborg and his collaborators Arthur Wahl and Joseph Kennedy chemically separated element 94 from uranium that had been bombarded with 16-MeV deuterons at Berkeley's 60" cyclotron. Not long after, Seaborg's group proposed the name plutonium for this new element. For the next year and a half, Seaborg's group performed numerous studies of both the nuclear and chemical properties of plutonium using only trace amounts of this new material. In order to study plutonium in its pure form, an effort was therefore made to produce a "macroscopic" amount of plutonium by irradiating many kilograms of natural uranium with neutrons produced by bombarding beryllium with accelerated deuterons. Cyclotrons at both Washington University and Berkeley were utilised for this production. After chemically extracting the plutonium from the uranium and fission products, on September 10, 1942, at the Metallurgical Laboratory of the University of Chicago, Boris Cunningham and Lewis Werner succeeded in producing the first sample of pure PuO_2 that was large enough to be weighed on a newly developed microbalance. The mass of the PuO_2 was determined to be 2.77 μg. This sample was preserved by sealing the platinum boat and oxide deposit inside a glass tube. The subsequent history of this sample is not known, but it was on display for a number of years at the Lawrence Hall of Science in Berkeley, CA.

At some point in the late 2000's, as a result of both financial and perceived radiation safety concerns, the Lawrence Hall staff decided to remove this sample from display. In 2008, a plastic box with a label on it stating "First sample of Pu weighed 2.7 μg" was found at the Hazardous Material Facility on the University of Berkeley's campus. It was assigned an EH&S sample number S338. There it could have been discarded as radioactive waste were it not for the knowledgeable eye of Phil Broughton, from UC Berkeley's Environmental Health and Safety Department. The paper trail documenting this sample's history had been lost and so the question was what could be done to establish its authenticity as Seaborg's plutonium?

In an attempt to authenticate this sample without opening the plastic box, we performed passive X-ray and gamma-ray analysis using both planar and coaxial high-resolution Ge detectors. A background-subtracted gamma-ray spectrum observed in approximately one day of counting is shown in Fig. 1. The object was found to contain ^{239}Pu. No other radioactive isotopes were observed. We carefully measured our gamma-ray detection efficiency for this sample and thus were able to quantitatively determine the mass of ^{239}Pu contained in the box to be 2.0±0.3 μg. These observations are consistent with the identification of this object being the 2.77-μg PuO_2 sample described by Glenn Seaborg and his collaborators as the first sample of ^{239}Pu that was large enough to be weighed [1].

2 Yield of the Trinity nuclear explosion

In 1945, the first nuclear explosive device was tested in Alamogordo, New Mexico. It was an implosion-type device using ^{239}Pu as the fissile material. Following the explosion, fission products and other materials contained in the "gadget" along with dirt and debris from the surrounding desert were caught up in the resulting fireball and heated to extremely high temperatures. When this material cooled and fell to the ground some of it was incorporated into a greenish glassy material now called Trinitite. We obtained four samples of Trinitite from Mineralogical Research Co. in San Jose, CA. These samples ranged in mass from 9 to 16 grams and were each counted using a planar Ge detector. Portions of the spectra observed from one of these samples are shown in Figs. 2 and 3.

[a]e-mail: ebnorman@lbl.gov

 DOI: 10.1051/epjconf/201612304001

As can be seen in these figures, we observed the characteristic ^{239}Pu gamma-ray lines at 52 and 129 keV along with the 662-keV line from the long-lived fission product ^{137}Cs. All other fission products would have decayed to undetectable levels (except for ^{90}Sr that does not emit any gamma rays). We carefully measured our gamma-ray detection efficiencies for the extended and unusual-shaped samples and then determined the activities of ^{239}Pu and ^{137}Cs contained in each one. Based upon these results, we then performed a very simple-minded analysis to determine the explosive yield of the Trinity device. To do so, we assumed the same probability for Cs and Pu to become incorporated in Trinitite. Due to high volatility of Cs, this should lead us to underestimate the yield. We also neglected the fission yield from the tamper of depleted uranium. This would also cause us to underestimate the yield. There is a very interesting interplay of both physics and chemistry involved in this analysis. The results from our four samples produce yields that range from 3.5 – 9.8 kilotons. These are all substantially lower than the actual yield of 21 kilotons, as could be expected from the type of analysis we carried out, but are at least the correct order of magnitude.

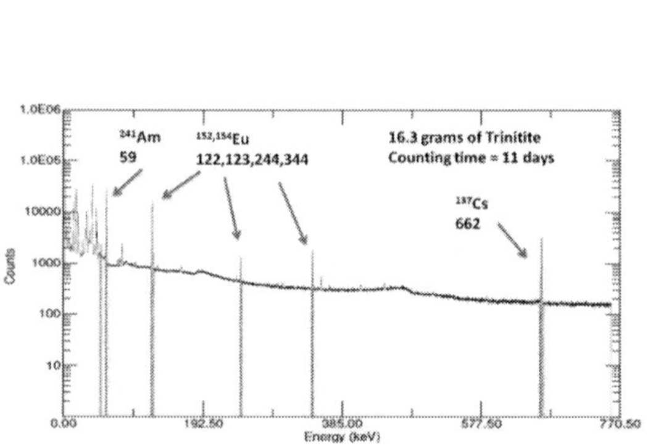

Figure 2: Energy spectrum observed from the 16.3-gram Trinitite sample in 11 days of counting. The 662-keV line from the decay of the long-lived fission product ^{137}Cs is clearly seen.

3 Beta-delayed gammas from fission

When a nucleus undergoes fission, the products emit both prompt and delayed characteristic gamma rays. The delayed gamma rays are emitted following beta decays, and therefore also have characteristic time scales associated with their emissions. These gamma rays can be used to identify fissionable material hidden in cargo containers, or distinguish between fissionable isotopes, and are useful for assay of nuclear fuels. They can also be potentially used in nuclear forensics to identify the average energy of neutrons producing fission and the original fissionable isotope.

However, for beta-delayed gamma-ray spectroscopy to be a useful tool, a complete set of measurements on all likely fissionable isotopes is needed. We, therefore, performed measurements of beta-delayed gammas following the thermal and 14-MeV neutron-induced fissions of ^{235}U and ^{239}Pu and also the fast-neutron induced fissions of ^{232}Th, ^{237}Np and ^{238}U. We used both high-resolution germanium detectors and low-resolution plastic scintillators to measure gamma energy and time spectra. We found distinctive gamma-ray intensity ratios for each sample in several time bins, which provide adequate data to identify the fissionable material present and can also provide information on the neutron energy [2, 3].

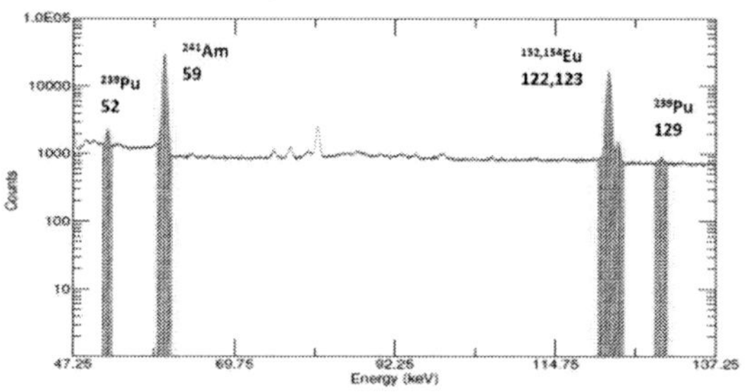

Figure 3: Low-energy portion of the spectrum observed from the 16.3-gram Trinitite sample. The 52- and 129-keV lines from the decay of ^{239}Pu are clearly seen.

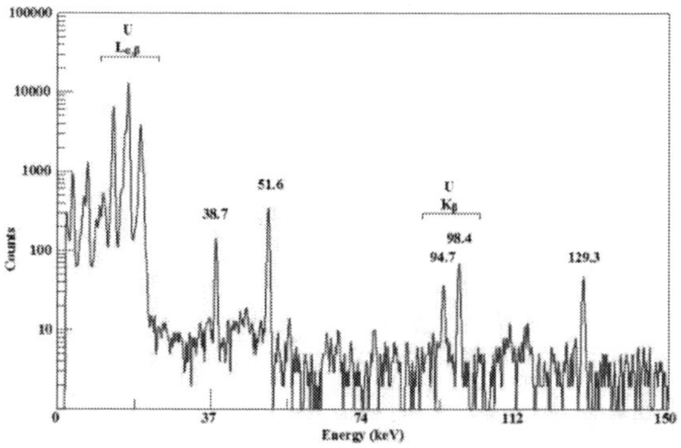

Figure 1: Background-subtracted spectrum observed in our planar Ge detector from Sample S338. All of the labeled peaks are X rays and gamma rays attributable to the decay of ^{239}Pu.

Figure 4 illustrates some of the differences we observed following the thermal-neutron induced fissions of ^{235}U (shown on top) and ^{239}Pu (shown below). Similar differences in gamma-ray line intensities were observed at a number of different energies from 5 minutes to 14 hours after fission. Using ratios of gamma-ray line intensities,

DOI: 10.1051/epjconf/201612304001

we showed that one can readily determine which isotope underwent fission. Similarly, one can use gamma-ray line intensity ratios to learn about the energy of the neutron that induced the fission. Figure 5 illustrates some of the differences we observed following the thermal and the 14-MeV neutron induced fission of ^{235}U.

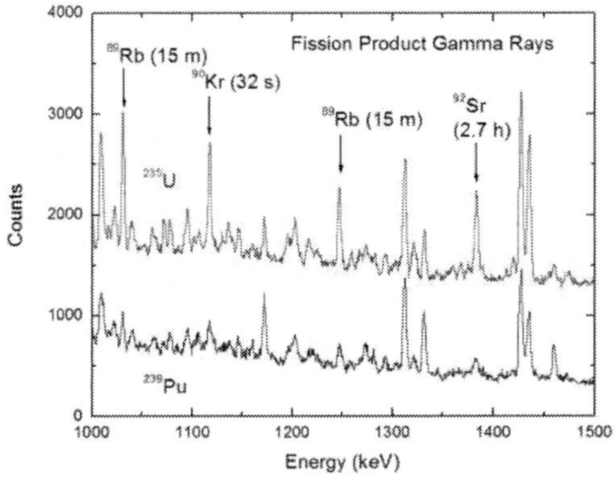

Figure 4: Arrows indicate lines from isotopes produced more strongly in ^{235}U fission than ^{239}Pu fission.

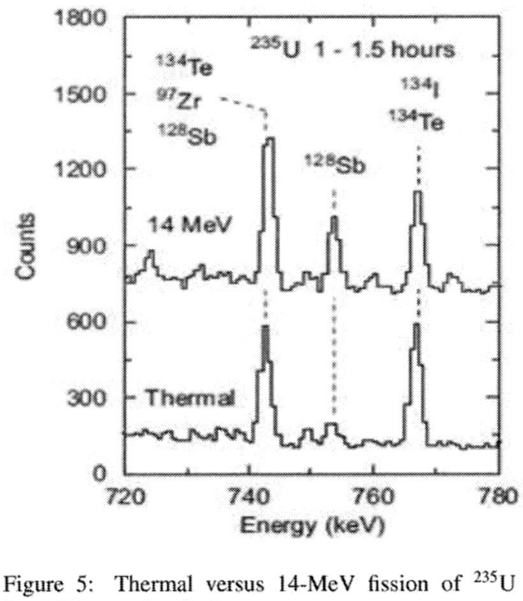

Figure 5: Thermal versus 14-MeV fission of ^{235}U at 1 – 1.5 hours after fission.

4 Fukushima fallout in the US

In order to search for possible worldwide distribution of radioactive contamination following the accident at the Fukushima Dai-ichi nuclear power plant in Japan, we collected rainwater samples in Berkeley, Oakland, and Albany, California and examined them for the presence of above normal amounts of radioactivity. Samples were collected from March 16 – March 26, 2011, by placing plastic containers outside in the Oakland and Berkeley hills and in Albany. Collection periods varied from a few to approximately 12 hours. Following collection, each sample was placed directly into a Marinelli beaker for gamma-ray counting. No processing of any kind was done to the rainwater samples before counting.

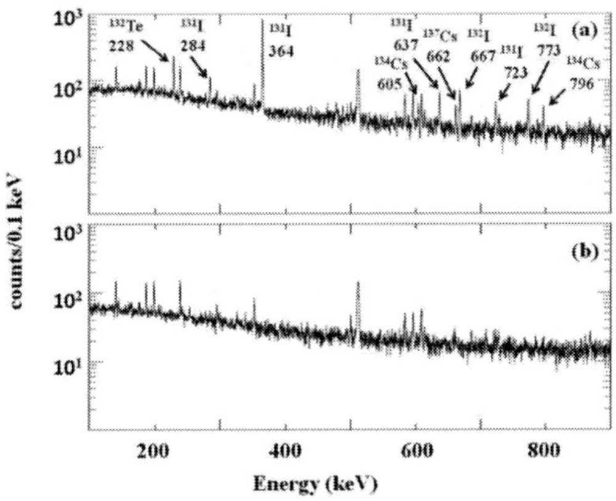

Figure 6: (a) A portion of the gamma-ray spectrum observed from counting 1 litre of rainwater collected in Oakland on March 18. (b) The same region of the spectrum collected from counting a 1-litre sample of Berkeley tap water. Both spectra represent the same counting period of 24 hours. Note that the gamma-ray peaks seen in the bottom spectrum do not originate in the tap water but are part of the natural background present in our laboratory environment.

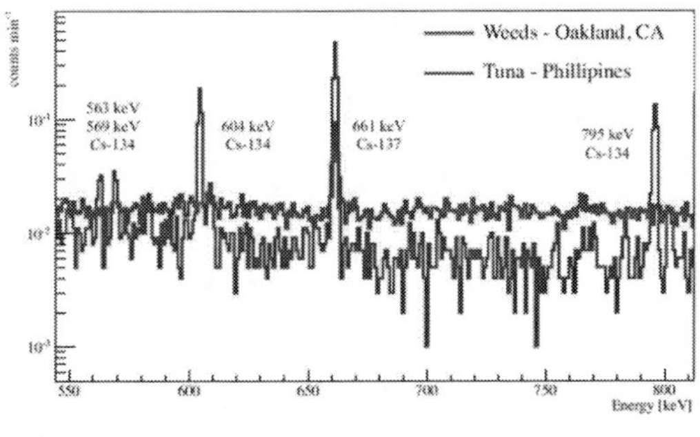

Figure 7: Comparison of γ spectra of local weeds collected in April 2011 in Oakland, CA along with a sample of Tuna from the Philippines purchased locally at a San Francisco Bay Area retail location.

As can be seen in Fig. 6, our spectra indicated the presence of ^{131,132}I, ^{132}Te, and ^{134,137}Cs. The levels we observed were low and were not a health risk to the public [4]. Subsequently we collected and counted numerous food samples of fish, seaweed, and dairy products being sold in the San Francisco Bay Area. Portions of the spectra we observed from samples of weeds collected in April, 2011 in Oakland, CA (that show ^{134,137}Cs from Fukushima contamination) and from Philippine tuna fish sold in Berkeley, CA in 2013 are shown in Fig. 7. We did not observe any evidence of Fukushima contamination in any of the food samples studied. We did, however, observe very low levels of ^{137}Cs in a number of food samples that probably originated from atmospheric nuclear weapons tests conducted in the 1950's and 1960's [5].

5 Conclusions

The measurements discussed in this paper illustrate how, in addition to its utility in fundamental studies in nuclear physics, gamma-ray spectroscopy can be used to study a number of interesting problems in nuclear forensics.

Acknowledgements

I wish to thank all of my collaborators in these projects. This work was supported in part by the U. S. Department of Energy National Nuclear Security Administration under Award Number DE-NA0000979 and by the U. S. Department of Homeland Security.

References

[1] E. B. Norman, K. J. Thomas, and K. E. Telhami, Amer. J. Phys. 83, 843 (2015).

[2] R. E. Marrs *et al.*, Nucl. Instrum. Meth. A 592, 463 (2008).

[3] A. Iyengar *et al.*, Nucl. Instrum. Meth. B 304, 11 (2013).

[4] E. B. Norman, C. T. Angell, and P. A. Chodash, PLoSONE 6(9):e24330 (2011).

[5] A. R. Smith *et al.*, J. Environ. Prot. 5, 207 (2014).

EPJ Web of Conferences **123**, 04002 (2016)
Heavy Ion Accelerator Symposium 2015
DOI: 10.1051/epjconf/201612304002

Searching for Dark Matter at the Stawell Underground Physics Laboratory

Phillip Urquijo[1,a]

[1] *School of Physics, The University of Melbourne, Parkville, Victoria 3010, Australia*

Abstract. facility to be built in 2016, located 1 km below the surface in western Victoria, Australia. I will discuss the status of the proposed SABRE experiment, which will be comprised of a pair of high purity 50-60 kg NaI crystal detectors with active veto shielding to be located in labs in the Northern and Southern Hemispheres respectively. I also discuss projects beyond SABRE, including directional dark matter detectors, which will be used to determine the origin of any true dark matter signals.

1 Introduction

An understanding of the nature of dark matter is one of the most important challenges faced by particle- and astrophysics today. About 30% of the total energy density in our Universe is believed to be in the form of relic massive particles interacting only feebly with ordinary matter. One experimentally allowed way for these particles to interact is via the weak interaction, similar to a neutrino. Direct detection of so-called WIMPs (Weakly Interacting Massive Particles) is possible by measuring the nuclear recoil energy following an elastic WIMP scattering off nuclei.

Over the past decade, several dark matter direct detection experiments have found intriguing anomalies, which may be due to WIMP interactions through the combined motion of the Earth in the solar system and in the Galaxy. The most striking of these results was from the DAMA/LIBRA experiment [1], which claimed 9σ evidence for annual modulation due to a dark matter hypothesis [2].

The Stawell Underground Physics Laboratory (SUPL) will be key in confirming or refuting the claims of DAMA/LIBRA, as host to the southern hemisphere experiment of the SABRE twin experiment. SUPL is the first deep underground laboratory in Australia, and the southern hemisphere. It will be host to dark matter experiments, and to other deep underground science research. In this proceedings I review the phenomena of dark matter annular modulation, the plans for SUPL, SABRE, and directional experiment R&D.

2 Dark Matter Observables

Dark matter searches are at the interface of particle-, nuclear- and astro-physics. In the case of WIMP dark matter, the differential nuclear recoil event rate with respect to energy due to dark matter interactions can be written as

$$\frac{dR}{dE}(E,t) = \frac{\rho_0}{m_\chi \cdot m_A} \cdot \int v \cdot f(\mathbf{v},t) \cdot \frac{d\sigma}{dE}(E,v) d^3v \quad (1)$$

where m_χ is the dark matter mass and $d\sigma/dE(E,v)$ its differential cross section [3]. The WIMP cross section, σ, and m_χ are the two observables of the experiment. The dark matter velocity v is defined in the rest frame of the detector and m_A is the target nucleus mass. The astrophysical parameters are the local dark matter density, ρ_0, and $f(\mathbf{v},t)$, which accounts for the WIMP velocity distribution in the detector reference frame.

Experimentally, both the energy dependence and the annual modulation of the dark matter signature are analysed. The WIMP-nucleus interaction rate is expected to modulate yearly due to the Earth's rotation around the Sun. The speed of dark matter particles in the Milky Way halo relative to the Earth is largest around June 2nd and smallest in December. Consequently, the number of particles able to produce nuclear recoils above the detectors' energy threshold is also largest in June. As the amplitude of the variation is expected to be small, the temporal variation of the differential event rate can be written as

$$\frac{dR}{dE}(E,t) \approx S_0(E) + S_m(E) \cdot \cos\left(\frac{2\pi(t-t_0)}{T}\right) \quad (2)$$

where t_0 is the phase expected at about 150 days and T is the expected period of 1 year. The time-averaged event rate is denoted by S_0, and the modulation amplitude by S_m.

3 Previous measurements

Owing to the very low expected event rate, about 10^{-2} events/kg/day for standard assumptions on the WIMP mass (100 GeV), experiments aiming at a direct dark matter search have very stringent background requirements [3]. Nuclear recoils induced by dark matter interactions

[a] e-mail: phillip.urquijo@unimelb.edu.au

are in an energy region heavily affected by environmental radioactive, seasonal, and cosmic background. The only way to adequately reduce the flux of cosmic ray-induced particles is to build the experiment deep underground.

For over a decade, the DAMA experiment (DAMA-NaI and DAMA/LIBRA), situated at Gran Sasso National Laboratory (LNGS), has been observing a low energy (2-6 keV_{ee} or keV electron equivalent) rate annular modulation in an array of high purity NaI(Tl) crystal detectors, shown in Fig. 1. This modulation occurs with a phase and period consistent with predicted dark matter interactions. Dark matter of similar mass and nuclear cross-section was also recently suggested by the CoGeNT[4], CDMS-Si, [5] and CRESST II [6] experiments. However, the dark matter interpretation of the DAMA/LIBRA modulation remains in question. Its tension with other experiments such as XENON [7], LUX [8] and KIMS [9] suggests the need for further investigation of the DAMA modulation. So far there are no firm explanations for the modulation other than a dark matter hypothesis. Although limits from large noble gas experiments, such as XENON and LUX, are very powerful in constraining dark matter, the WIMP interaction processes they are sensitive can differ to DAMA/LIBRA, and therefore cannot rule it out in a model independent way.

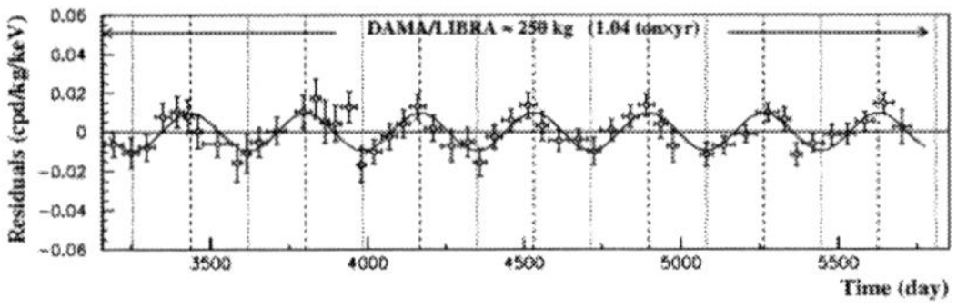

Figure 1. The annual modulation of the low energy event rate in the DAMA/LIBRA experiment. Figure taken from Ref. [2].

In general, the differential WIMP-nucleus cross section, $d\sigma/dE$ can be written as the sum of a spin-independent (SI) contribution and a spin-dependent (SD) one,

$$\frac{d\sigma}{dE} = \frac{m_A}{2\mu_A^2 v^2} \cdot \left(\sigma_0^{\rm SI} \cdot F_{\rm SI}^2(E) + \sigma_0^{\rm SD} \cdot F_{\rm SD}^2(E)\right) \quad (3)$$

where F_i are the wimp-nucleon form factors, σ_{0i} are the cross sections at zero momentum transfer, m_A is the target mass, and μ_A is the WIMP-nucleus reduced mass. The form factors can be determined in nuclear shell model calculations. The spin independent cross sections are typically simple functions of the nucleon number, while the spin dependent cross sections have most recently been determined using chiral effective-field theory and depend on isotope. A summary of the current status of dark matter searches is shown in Fig. 2 for spin-independent limits..

There are other scattering processes that may be allowed by the current experimental data. Various models allow dark matter scattering off electrons. For example, sub-GeV dark matter particles could produce detectable ionisation signals. If new forms of couplings are introduced to mediate dark matter-electron interactions, further models become viable. Assuming an axial-vector coupling, dark matter-lepton interactions dominate at tree level and cannot be probed by dark matter-baryon scattering. Models such as kinematic-mixed mirror dark matter [11] also predicts interactions with atomic electrons.

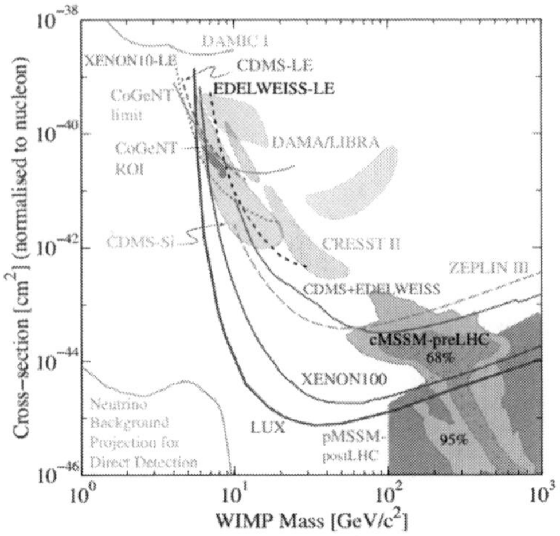

Figure 2. WIMP cross sections (normalised to a single nucleon) for spin-independent coupling versus mass, taken from Ref. [10].

Although alternative models may still be able to reconcile DAMA/LIBRA signals with other experimental results, a model-independent test using ultra high purity NaI(Tl) crystal detectors is necessary to confirm or refute the DAMA/LIBRA dark matter claim. The SABRE collaboration proposes to test the DAMA signal in a dual site approach at LNGS (the site of DAMA/LIBRA), and in Stawell. The experiment will consist of highly pure NaI(Tl) crystals in an active liquid scintillator veto to tag and reduce the potassium background. Two identical detectors will be made, for the two sites. This approach allows for substantial mitigation of season modulation effects, particularly seasonal temperature variations.

4 Low background physics

The allowed cross section of a WIMP interaction is extremely low. In order to identify unambiguous signs of dark matter particles, ultra-low background experimental conditions are required. Background from γ-rays, and neutrons must be mitigated to great extent. The dominant radiation from gamma-decays originates from uranium and thorium chains, as well as common isotopes, ^{40}K, ^{60}Co and ^{137}Cs present in the surrounding materials. The ^{232}Th and ^{238}U chains have a series of alpha and beta decays accompanied by the emission of several gamma rays with energies from tens of keV up to 2.6 MeV. Many of the possible interactions of gamma rays with matter can deposit energy in the target medium in the range of a few keV, which is the sensitive region of detectors like DAMA/LIBRA. To stop γ activity from outside of the experimental setup, large water tanks are often em-

DOI: 10.1051/epjconf/201612304002

ployed, although not ubiquitously. They provide homogenous shielding, and satisfy most background requirements. This demands large experimental halls, many times greater than the dimensions of the active detector region.

Neutrons can interact with the nuclei in the target via elastic scattering, mimicking the nuclear recoil signature of a WIMP interaction. The most challenging case is elastic scattering, as opposed to in-elastic scattering, which is usually accompanied by gamma emission that can be tagged by inducing scintillation in multiple crystals. Most cosmogenic neutrons are produced due to spallation reactions of muons on nuclei in the experimental setup or in the surrounding rock. While spallation neutrons can be several GeV in energy, subsequent interaction with the detector material reduces the energies to MeV ranges, which can produce nuclear recoils in the dark matter search regime. The most powerful way of moderating the flux of this background is to place the detectors as deep as possible. Radiogenic neutrons must be controlled with good material selection, and a good screening facility.

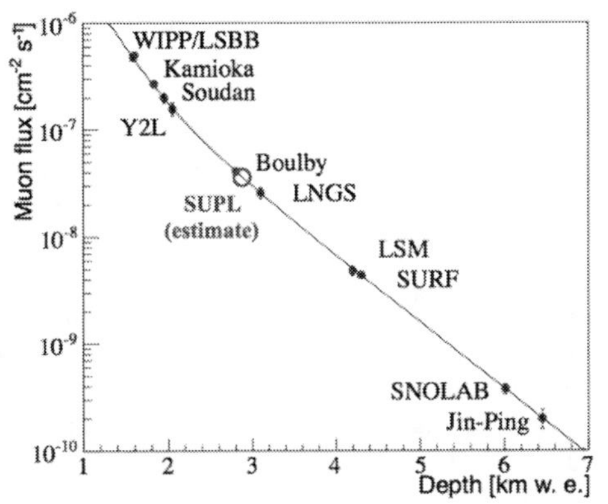

Figure 3. Muon flux as a function of depth in km of water equivalent for various underground laboratories. The SUPL depth and muon flux is approximate pending precision results. The parameterisation (solid line) is taken from Ref [14].

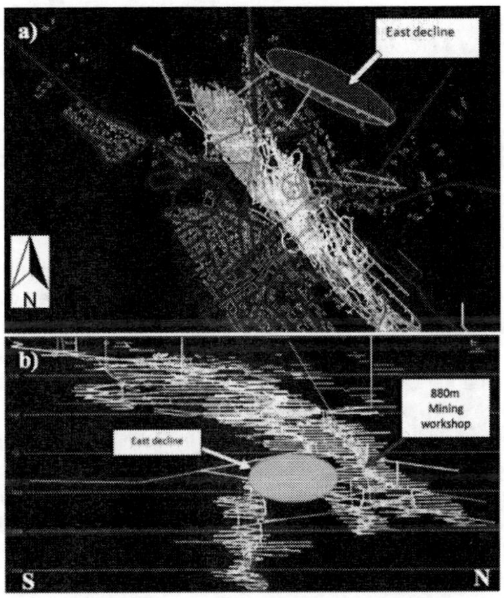

Figure 4. The proposed location of SUPL in the Stawell Gold Mine: a) Topological view of Stawell and the mine tunnels underneath, b) side view of the tunnels on the North-South projection. The laboratory will be located near the east decline.

5 Stawell Underground Physics Laboratory

An Australian dark matter program has recently been established, to provide unique insight into this puzzle, and to make important contributions to the field of low background physics. The proposed laboratory will be located at the Stawell Gold Mine, Stawell, Victoria, Australia. The mine is 250 km west of Melbourne. The Stawell Gold Mine is a decline mine with a helical access road descending the main ore body with many large caverns dug into the ore body along the road. It is a dry basalt rock mine with a flat overburden, and reaches a depth of 1.6 km. The mine is still active.

The location and the nature of the mine make it ideal for dark matter experiments. We have identified a site at a depth of 1025 m to host the laboratory. Given the high density of basalt, 2.86 g/cm^3, the site has approximately 2900 m water equivalent of overburden, similar to the average depth of the Gran Sasso National Laboratory in Italy (Fig. 3). It is situated near the east decline, where there are no mining tunnels overhead, shown in Fig. 4. The laboratory is planned to contain a dark matter experiment and an integrated underground laboratory facility for hosting other research.

A series of preliminary characterisation measurements have been carried out at a depth of 880 m to estimate the conditions of the laboratory. These are being repeated near the nominal laboratory site. The neutron flux was found to be approximately 7×10^{-6} n s^{-1} cm^{-2} for fast neutrons (0.5 MeV and above) and 2×10^{-6} n s^{-1} cm^{-2} for slow neutrons (< 0.5 MeV), however a dedicated program of precise neutron energy spectrum measurements is underway to inform shielding design. The upper limit on the γ ray flux for energies below 3 MeV is 2.5 γ s^{-1} cm^{-2}. The surrounding rock contained approximately 0.6 ppm ^{238}U and 1.6 ppm ^{232}Th. The ^{40}K concentration was found to be as low as 20 Bq/kg. These conditions have been found to be within acceptable tolerances as defined by LNGS, *i.e.* similar to Gran Sasso. The muon, slow-neutron, fast-neutron and gamma ray fluxes are currently being measured with high precision to inform detector shielding design. A new muon detector setup has been built in Melbourne to ensure low noise, and maximum muon purity. The first results from the upgraded muon detector are due in early 2016. The radon levels are acceptable when ventilation is provided to the experiment area, and will be further controlled within each experiment.

Preliminary plans for the design of SUPL have been drafted, and the engineering company responsible for leading the construction has been tendered. The laboratory space will be built during 2016, for delivery in early 2017. The design is to have a main cavity of 35-40 m (length) × 10 m (width) × 10 m (height), with a loading area of 15 m (length), and a class 1000 clean area of 25 m (length), shown in Fig. 5. It will house a clean ante room and washing area, a separate high grade clean room, and two low background screening facilities. The design of the laboratory will allow for a possible expansion space in the future.

SUPL is planned to be a multi-use underground laboratory. Its flagship will be the SABRE dark matter detector, but other uses will include nuclear screening, astrobiology, and as a host for research and development of new underground experiments in dark matter and neutrino physics.

6 SABRE

The SABRE (Sodium iodide with Active Background REjection) experiment seeks to conduct a direct test of DAMA/LIBRA through a number of innovations. The key features are the development of ultra-high purity NaI(Tl) crystals and the rejection of the residual ^{40}K background (among others) using an active veto. High-purity crystals have been developed at Princeton. The active veto has been designed with a liquid scintillator vessel.

6.1 Experiment plan

The final SABRE setup will be two identical experiments with 50-60 kg of NaI crystals. A prototype, or commissioning detector will be conducted with a small 1-2 kg crystal. Each crystal will be enclosed in a self-contained module consisting of the crystal itself and two low-radioactivity, high quantum efficiency photomultiplier tubes. The phototubes are optically coupled to opposite sides of the crystal. These components are then sealed in a light and air tight low radioactivity metal container. The crystal detectors are placed inside of a cylindrical liquid scintillator, 1.5 m diameter and 1.5 m length, detector that is shielded from environment background with passive shielding. The passive shielding is still be designed for the respective locations at Gran Sasso and Stawell. The liquid scintillator detector is able to reject internal radioactivity as well as external γ ray backgrounds. The scintillator vessel will contain approximately 2 tonnes of linear alkylbenzene (LAB). Approximately twenty 20.2 cm Hamamatsu R5912 PMTs will be placed at the ends of the cylindrical vessel.

The most powerful, and novel aspect to the SABRE experiment is the dual location approach. By placing one detector in the southern hemisphere at Stawell (37°03′ S) and one in the northern hemisphere at Gran Sasso (42°28′ N), the seasonal thermal impact on background to dark matter annual modulations can be constrained. Dark matter candidates from the galactic halo wind will however provide consistent, in phase, signatures in the two locations. Each detector will use the same technology and detection approach, thereby reducing model dependence.

6.2 Active veto concept

Sodium Iodide crystals, such as those used in DAMA/LIBRA contain traces of ^{40}K (a 13 ppb average is reported). When ^{40}K decays to ^{40}Ar, it produces a 1.46 MeV γ-ray at the same time as a possible 3 keV X-ray/Auger electron. escapes the detector, the 3 keV deposition will become a source of background in the primary energy region of interest. Simultaneous, separate detection of these two signals, however, can be taken as a signature of ^{40}K decay and can be used to reject such background events. DAMA/LIBRA suppressed some of the ^{40}K background by rejecting multi-hit events in which more than one crystal sees an event. However, this approach was inefficient, as evidenced by the ~ 3 keV$_{ee}$ peak in their energy spectrum.

In the SABRE design, a dedicated detector surrounds the crystal modules, which can be used to detect radiation leaving the dark-matter sensitive volume. The outer detector is filled with liquid scintillator, LAB, and provides an active veto for radioactive background events such as ^{40}K decay. An illustration of the principle is shown in Fig. 6.

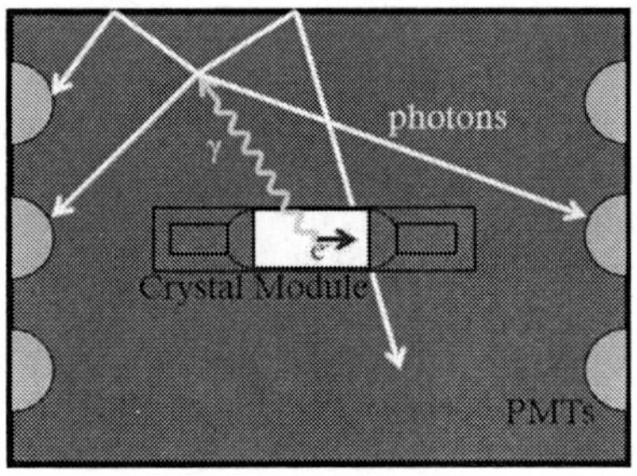

Figure 6. Illustration of the veto concept, demonstrating the detection of a ^{40}K decay X-ray/Auger electron and corresponding 1.4 MeV γ ray. The active veto observes the γ ray in the liquid scintillator material and the events are rejected.

6.3 Crystal production

The most important part of the experiment is the reduction of ^{40}K contamination in the scintillator crystals. The Princeton group partnered with Sigma-Aldrich to produce NaI powder with lower background than previously achieved. The primary aim has been to reduce ^{40}K levels below the 13 ppb seen by DAMA/LIBRA. There is also concern for background from U, Th, and Rb, which all contribute to the flat background in the energy spectrum. Sigma-Aldrich has made a high purity "Astro-grade" powder with a reported K level of 3.5 ppb. This powder has been used to grow test crystals, and is the nominal choice for the final full scale experiment.

DAMA/LIBRA has a single interaction rate of ~1 cpd/kg/keV$_{ee}$, with reported crystal radioactivity levels of 13 ppb nat K (average) and ~1-10 ppt for U and Th. The SABRE crystals are currently being grown from the "Astro-grade" powder. The scintillation yield and optical properties of the crystal are also very important: they

DOI: 10.1051/epjconf/201612304002

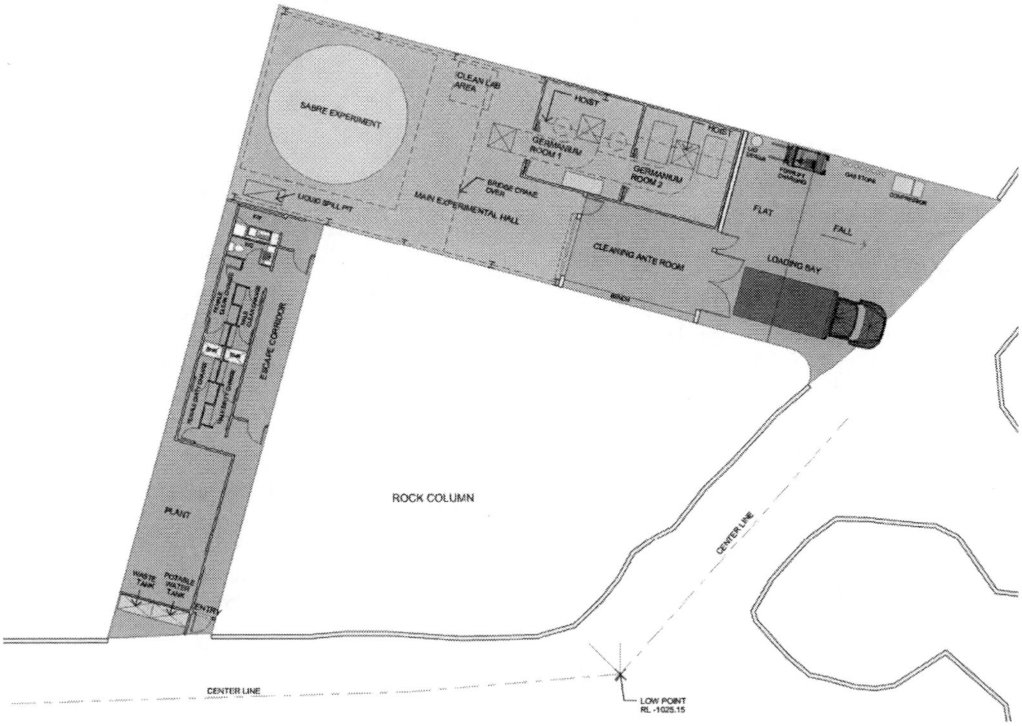

Figure 5. Conceptual design for the SUPL laboratory, located 1025 m below the surface. The dimensional units are in metres. The photographs are taken from other laboratories and are used for illustration.

are directly related to the detected signal amplitude, and thus to the achievable energy threshold of the experiment. Large crystal masses are also necessary for the experiment to achieve a high dark matter sensitivity. Due to light attenuation effects, longer crystals decrease the light yield, and crystals with large cross-section are preferred. We expect to grow NaI(Tl) crystals around 10.2 cm in diameter and 25.4 cm long with a total mass of 8-10 kg each. It is expected that the SABRE crystals will have much less than 1 count per day per kg.

The dark matter interpretation of the DAMA/LIBRA (or SABRE) modulation signal depends on the scintillation efficiency of NaI(Tl) for sodium and Iodine recoils relative to that produced by WIMP scattering interactions. Nuclear recoils in NaI(Tl) tend to produce less scintillation light than electron recoils with the same energy deposition. The relative ratio is usually referred to as the quenching factor. DAMA reports a quenching factor of 0.3 for sodium and 0.09 for iodine recoils, by exposing the NaI(Tl) crystals to a ^{252}Cf source. A more recent analysis by members of SABRE have found contradictory results that change in WIMP interpretation [12]. They use a source of low-energy, pulsed neutrons and combine time-of-flight and pulse-shape-discrimination techniques to perform a very precise measurement of the Na-recoil quenching factors. The results are shown in Fig. 7. The implications for DAMA/LIBRA are that the observed modulation signal occurs in the energy window of 13 keV to 32 keV instead of the 7 keV to 20 keV as previously quoted, shifting the WIMP mass interpretation to higher mass values.

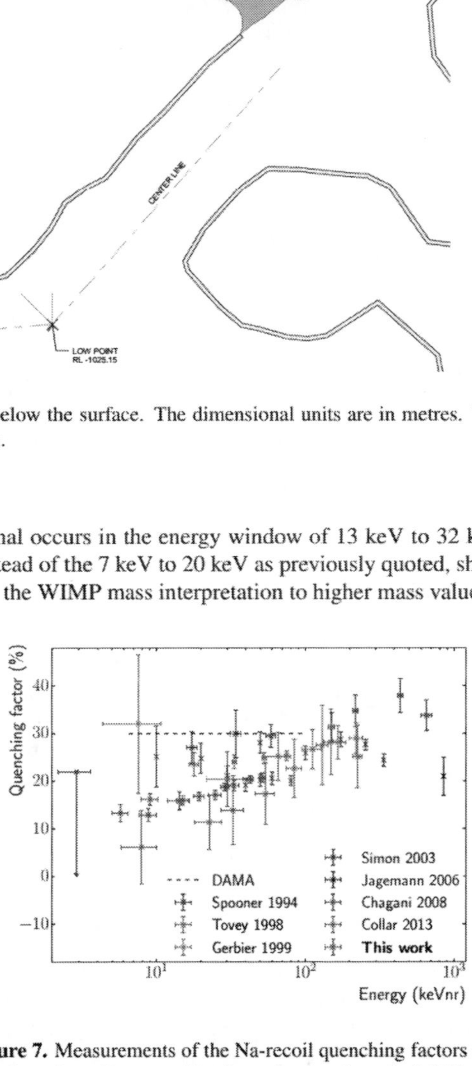

Figure 7. Measurements of the Na-recoil quenching factors as a function of recoil energy. The figure is taken from Ref. [12], and the most recent measurement shown with red markers.

6.4 Measurement expectations

An estimate of the background for a dark matter measurement has been performed using Monte Carlo simulation (MC) of a 2 kg crystal and assuming DAMA-level background in the crystal [13]. This simulation was used to demonstrate the veto detector's rejection efficiency. It was found that of all ^{40}K events depositing 1-4 keV energy in the crystal, nearly 90% are rejected by the veto or coincidence with other crystals, shown in Fig. 8. It is expected that the SABRE background level would be about two times less than the DAMA/LIBRA flat singles rate. With the conservative predicted background rate of 0.4 cpd/kg/keV$_{ee}$, described in Ref. [13], MC studies show that it would take SABRE 3 years to refute or confirm the DAMA/LIBA modulation signal with $> 3\sigma$ confidence level. The low radioactivity background achieved by SABRE should also allow it to probe new frontiers in WIMP dark matter searches.

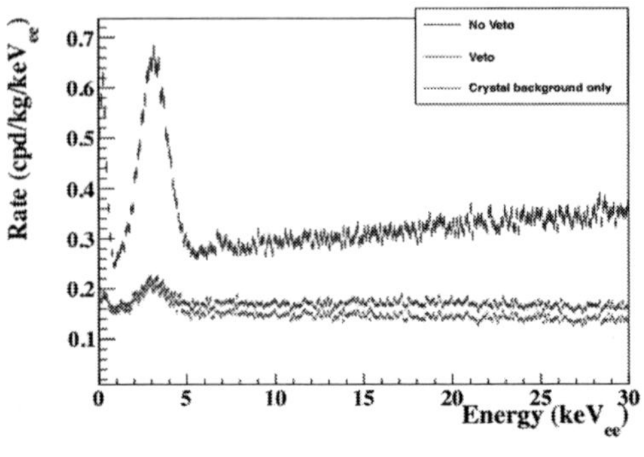

Figure 8. Estimated background in SABRE based on Geant4 Monte-Carlo simulation.

7 Directional Experiment Prospects

A positive signature from SABRE must be accompanied with a search program to further elucidate the nature and source of the dark matter signature. Directional dark matter detectors offer an excellent way to achieve this goal. The principle of direct detection is to measure both the energy and the trajectory, of a recoiling nucleus following a WIMP scattering process. The double differential spectrum is dependent on the WIMP velocity distribution, and hence the Milky Way halo model. The WIMP event distribution is expected to present an excess in the direction of motion of the Solar system, which happens to be roughly in the direction of the constellation Cygnus. The WIMP induced recoil distribution presents a dipole feature in galactic coordinates, whereas the background distribution is expected to be isotropic in the galactic rest frame. Stawell sits in a geographically advantageous location for directional dark matter studies, as the angle of inclination to the WIMP wind oscillates with the greatest amplitude of any underground laboratory (Fig. 9). This means the experimental signature will be the most pronounced.

The experimental signature of direct detection techniques is the recoil track produced in the dark matter interaction. This provides information on the ionisation density as a function of position, on the range and eventually on

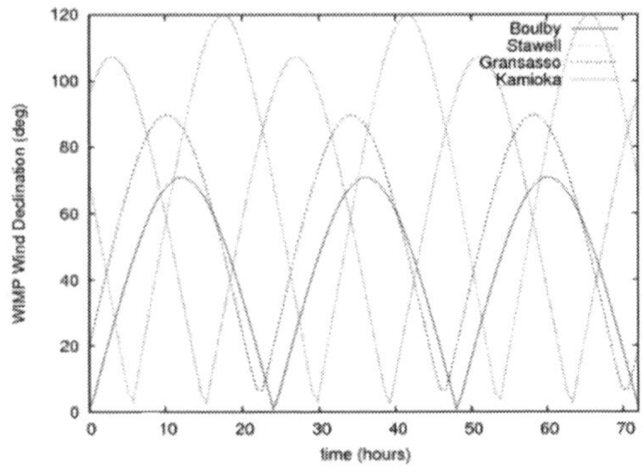

Figure 9. Declination angle to the WIMP wind (Cygnus).

the direction of the recoiling nuclei. The most promising strategy for directional searches is the use of low pressure gas, leaving tracks of 1-2mm in length. The most successful direct detection experiment is DRIFT-II, operated at Boulby [15]. Its volume is 0.8 m^3, and is filled with a low pressure mixture of CS_2:CF_4:O_2 gas. It contains 32.2 g of fluorine in the active volume for WIMP interactions. Research is ongoing with the CYGNUS collaboration to adapt this technology to very large scales, up to 64 m^3, and is a candidate project for SUPL in the near future.

8 Summary

SUPL will be a major new facility for Australia and for global underground science projects, to be constructed in 2016/2017. The flag-ship experiment, SABRE, aims to test the claims of the DAMA/LIBRA collaboration in a model independent way using sites in the north and south hemispheres. SUPL also offers opportunities for directional dark matter studies, which will be the only way to elucidate the nature of dark matter once found.

References

[1] R. Bernabei *et al.* [DAMA Collaboration], Nucl. Instrum. Meth. A **592**, 297 (2008).
[2] R. Bernabei *et al.* [DAMA Collaboration], Eur. Phys. J. C **56**, 333 (2008).
[3] T. M. Undagoitia and L. Rauch, J. Phys. G **43**, 013001 (2016).
[4] C. E. Aalseth *et al.* [CoGeNT Collaboration], Phys. Rev. Lett. **106**, 131301 (2011).
[5] R. Agnese *et al.* [CDMS Collaboration], Phys. Rev. Lett. **111**, 251301 (2013).
[6] G. Angloher *et al.*, Eur. Phys. J. C **72**, 1971 (2012).
[7] E. Aprile *et al.* [XENON100 Collaboration], Phys. Rev. Lett. **109**, 181301 (2012).
[8] D. S. Akerib *et al.* [LUX Collaboration], Phys. Rev. Lett. **112**, 091303 (2014).
[9] K. W. Kim *et al.*, Astropart. Phys. **62**, 249 (2014).
[10] K. A. Olive *et al.* [Particle Data Group Collaboration], Chin. Phys. C **38**, 090001 (2014).

[11] R. Foot, Phys. Rev. D **90**, 121302 (2014).
[12] J. Xu *et al.*, Phys. Rev. C **92**, 015807 (2015).
[13] E. Shields, J. Xu and F. Calaprice, Phys. Procedia **61**, 169 (2015).
[14] D. Mei and A. Hime, Phys. Rev. D **73**, 053004 (2006).
[15] J. B. R. Battat *et al.* [DRIFT Collaboration], Phys. Dark Univ. **9-10**, 1 (2015).

EPJ Web of Conferences **123**, 04003 (2016) DOI: 10.1051/epjconf/201612304003
Heavy Ion Accelerator Symposium 2015

Search for bound-state electron+positron pair decay

F. Bosch[1], S. Hagmann[1], P.-M. Hillenbrand[1], G. J. Lane[2], Yu. A. Litvinov[1,3,a], M. W. Reed[2,b], M. S. Sanjari[1], Th. Stöhlker[1,4,5], S. Yu. Torilov[6], X. L. Tu[1,3,7], and P. M. Walker[8]

[1] *GSI Helmholtzzentrum für Schwerionenforschung, Planckstraße 1, 64291 Darmstadt, Germany*
[2] *Department of Nuclear Physics, Research School of Physics and Engineering, The Australian National University, Canberra, ACT 2601, Australia*
[3] *Max-Planck-Institut für Kernphysik, Saupfercheckweg 1, 69117 Heidelberg, Germany*
[4] *Helmholtz-Institut Jena, 07743 Jena, Germany*
[5] *Institut für Optik und Quantenelektronik, Friedrich-Schiller-Universität Jena, 07743 Jena, Germany*
[6] *St. Petersburg State University, Ulyanovskaya str. 1, Petrodvorets, 198504 St. Petersburg, Russia*
[7] *Institute of Modern Physics, Chinese Academy of Sciences, Lanzhou 730000, China*
[8] *Department of Physics, University of Surrey, Guildford, Surrey GU2 7XH, United Kingdom*

Abstract. the heavy ion storage rings coupled to in-flight radioactive-ion beam facilities, namely the ability to produce and store for extended periods of time radioactive nuclides in high atomic charge states, for the search of yet unobserved decay mode – *bound-state electron-positron pair decay*.

1 Introduction

The question on whether nuclear halflives are fundamental constants or can be modified by external manipulations was asked already at the beginning of nuclear physics [1]. Only tiny effects of below about one percent were found, see review [2]. Recently, a 1.5% change in the electron capture decay rate of ^{7}Be implanted in C_{60} clusters was observed [3]. Unconfirmed and disputed claims exist on the dramatic acceleration of α-decay in nuclei implanted into metals at cryogenic temperatures [4]. In all these cases, the results were explained using the modifications of electron densities, which is one of the main quantities needed to describe decay processes involving bound electrons [5].

Even with this scarce experimental evidence, it has been observed that high atomic charge states can dramatically modify nuclear decay properties [6–8]. A striking example is ^{163}Dy. As a neutral atom it is stable. However, when fully-ionised its halflife becomes merely $T_{1/2}(^{163}\mathrm{Dy}^{64+}) \sim 33$ years [9]. In the case of ^{187}Re, removing all orbital electrons reduces the half-life by nine orders of magnitude [10]. Both nuclides decay via *bound-state β decay* in which the decay electron occupies one of the bound orbitals instead of being emitted to the continuum. This decay mode is a time-mirror of the electron capture decay. The process is marginal in neutral atoms, but it can be significant if atoms are highly-ionised.

It is immediately obvious that *orbital electron capture* and *internal conversion* decays are disabled in fully-ionised atoms. In such cases weak decay branches like *positron emission* (if energetically allowed) or γ *de-excitation* can be investigated. For instance, conversion coefficients of nuclear isomers [11–13], were measured in fully ionised ^{144}Tb, ^{149}Dy and ^{151}Er isomeric states. Investigations of decay probabilities as a function of atomic charge states reveal interesting results. Counterintuitive results were obtained in electron capture of one- and two-electron systems in ^{122}I, ^{140}Pr and ^{142}Pm ions [14–16], where the rate is by about 50% larger in the ion with one bound electron as compared to the ion with two electrons. These results can be explained by taking into consideration the conservation of the total nucleus plus leptons angular momentum [17, 18]. Due to the energetic blocking of the K-shell internal conversion, the decay rate of the 35.5-keV first excited state in $^{125}_{52}$Te is increased by 300% and 640% for 47+ and 48+ charge states as compared to the value known for neutral atoms [19]. A new decay mode *bound-state internal conversion* was observed for few-electron Fe isomers [20, 21]. Here the isomer de-excitation energy is transferred to a bound electron which is then excited to a higher, unoccupied atomic level. Increased half-life of the isomeric state in ^{192}Os with a single bound electron allowed for benchmarking conversion coefficient estimations [22]. Studying the decays of isomers in highly-charged ions is important not only for investigations of weak decay branches or for testing of theoretical calculations, it is essential also for nuclear astrophysics [23], where the pathways of nucleosynthesis may dramatically be altered if low-spin long-lived excited states are present [24]. In particular this relates to $0^+ \to 0^+$ ($E0$) transitions which are strongly suppressed in fully-ionised atoms.

[a] e-mail: Y.Litvinov@gsi.de
[b] e-mail: Matthew.Reed@anu.edu.au

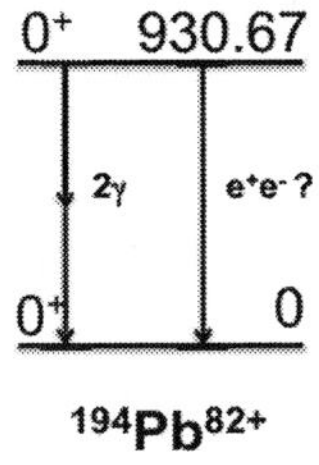

Figure 1. The decay scheme of the first excited state in ^{194}Pb.

2 Bound-state electron-positron pair decay

Since a photon is a boson with unit spin, the $0^+ \to 0^+$ transitions via a single photon emission are forbidden. In neutral atoms such decays proceed via internal conversion or, if the transition energy is larger than two times the rest mass of electron $m_{e^-} = 511$ keV, via electron-positron pair decay, $e^- + e^+$. Furthermore, double gamma-ray emission is possible [25–28].

In the completely ionised atom, decay via internal conversion is impossible and, if $e^- + e^+$ is energetically disabled, the half-life of the state can dramatically be extended since the double γ-ray decay is slow (second order in α, $\propto \alpha^2$). This can have dramatic consequence, for instance, for rapid proton-capture nucleosynthesis process running along the $N = Z$ line, where many even-even nuclei with first excited 0^+ states are known.

In this paper we propose to search for a *bound-state electron-positron pair decay*, the yet unobserved decay mode. $E0$ decays with transition energies below the 1022-keV pair production threshold exist in neutron-deficient lead nuclei, where low-lying 0^+ states are a consequence of nuclear shape coexistence [29].

One of the prominent candidates for this study is the magic $^{194}_{82}\mathrm{Pb}^{82+}_{112}$. The decay scheme of the first excited state in ^{194}Pb is illustrated in Fig. 1. The energy of the 0^+ state is $E(0^+) = 930.67$ keV which is known from precision spectroscopy investigations [30]. This energy is below 1022-keV threshold for $e^- + e^+$ decay. However, the binding energy of the last K-electron is $B(K) = 101.336$ keV [31, 32], which makes the new decay mode, bound-state $e^- + e^+$ decay, energetically allowed. In this case the electron shall occupy the vacant K-shell thus saving ~ 100 keV energy. The excess energy of about 10 keV is taken away by a monochromatic positron, which may appear as a forward-emitted particle in the laboratory frame.

3 Experiment

It is proposed to employ a heavy-ion storage ring coupled to an radioactive-ion beam facility for the search of the bound-state $e^- + e^+$ decay. Ions of ^{194}Pb produced in the first excited state can be stored as fully-ionised atoms in a ring. The quantum state in which ^{194}Pb are stored will continuously be monitored by the time-resolved Schottky mass spectrometry (SMS) [33, 34]. Nuclides with different masses are resolved by their revolution frequencies. The noise power of the Fourier-transformed signal directly relates to the number of stored ions. The SMS is sensitive to single particles [35] and has been employed for the search of long-lived isomers [36–39] as well as new isotopes [40, 41]. With new resonant detectors, time resolution of a few milliseconds can be envisioned [42–44]. The mass resolving power of the SMS is about 700 000 [45].

The decay in the ring is characterised by the disappearance of the signal at the revolution frequency of the parent ion and the correlation in time to the appearance of a signal at the frequency corresponding to the daughter ion [46]. The change in frequency reflects the change in the mass-over-charge ratio between the parent and the daughter ions. The details on the single-particle decay spectroscopy in storage rings can be found in [47].

Isomeric states decaying by double gamma-ray emission will not change the charge state and the daughter ion is still $^{194}\mathrm{Pb}^{82+}$ in its ground state. The bound-state $e^- + e^+$ decay results in $^{194}\mathrm{Pb}^{81+}$, which has a distinctly different revolution frequency. There is however a source of background, which is due to atomic pick-up of an electron from the rest-gas yielding the same daughter ions. If electron cooling of the stored ions [48] is employed, then also the recombination with the cooler electron has to be considered. Alternatively to electron cooling, the ions can be cooled stochastically [49, 50] or the ring can be operated in the isochronous ion-optical mode [45, 51, 52].

To perform first feasibility studies, one could use the operating facility at GSI Helmholtzentrum for Heavy Ion Research in Darmstadt, Germany, which is a combination of the fragment separator FRS [53] and the experimental storage ring ESR [54]. One could, for instance, study the most probable way to produce the first 0^+ in $^{194}\mathrm{Pb}^{82+}$. Direct production at the FRS or the production directly in the ESR shall be considered. In-ring α-decays were suggested for the ESR to study the electron screening effects [55], though no systematic investigations were performed so far (see [56] for more details). The α-decay branchings $^{198}_{84}$Po populating the states in $^{194}\mathrm{Pb}^{82+}$ are known [30], and the in-ring α-decay decay might be an efficient method to produce the state of interest. Concerning the background, one way to remove it could be the extraction and storage of ions in a Penning trap coupled to the ESR, HITRAP [57]. In this case, a clean detection method would be to measure 11-keV positrons from the decay. A possibility to distinguish the atomic electron capture from the isomer de-excitation decays on the basis of the distributions of the ion recoils (see [47]) shall be investigated as well. Monochromatic positrons emitted by fast projectiles is a favourable background-free detection method. Detection of ~ 11-keV positrons is feasible with an electron-spectrometer installed at the ESR [58, 59]. The concept of detecting positrons is described in detail in [60].

With higher kinetic energies the probability to pick-up electrons rapidly decreases. Therefore, another facility for the experiment could be the Facility for Antiproton and Ion Research, FAIR, which is under construction in Darmstadt [61]. There, the new fragment separator Super-FRS [62] and the isochronous Collector Ring [63] would allow for such kind of experiment at a much higher energy (740 MeV/u) than presently possible at GSI (400 MeV/u). We note that the High Energy Storage Ring, HESR, will allow for storing ion beams to energies of up to 5 GeV/u [64, 65]. However, the acceleration time of several ten seconds is not suitable for the present experiment. Similar to FAIR, such experiment could be considered for the High Intensity Accelerator Facility, HIAF, which is in a planning stage in China [66].

The highest primary beam intensities are presently available at RIKEN Nishina Center in Wako, Japan. Here an isochronous storage ring Rare RI Ring (R3) [67, 68] coupled to the BigRips fragment separator was commissioned in 2015. Due to the DC-nature of the RIKEN main cyclotron, individual ions are identified in flight and are injected into the R3 at about 400 MeV/u. This energy is the same as can be achieved at GSI. However, dependent on the production rate of the 0^+ state of interest, the facility in RIKEN can be superior for this experiment.

4 Summary

The bound-state $e^- + e^+$ decay is energetically allowed for the first 0^+ excited state in fully-ionised $^{194}Pb^{82+}$. The observation of this decay mode can experimentally be addressed at heavy-ion storage rings coupled to the radioactive-ion production facilities [69]. Such experiments, however, are very difficult at present but can be planned at the new generation radioactive-ion beam facilities like FAIR or HIAF.

In memorium

This short paper is dedicated to George Dracoulis, our friend who was inspired by the application of the heavy-ion storage rings to isomer studies.

Acknowledgements

This work is supported in part by the Helmholtz-CAS Joint Research Group (HCJRG-108), the Australian Research Council (DP120101417 and FT100100991), and the UK-Science and Technology Facilities Council.

References

[1] E. Rutherford and F. Soddy, J. Chem. Soc., Trans. **81**, 837 (1902).
[2] G. T. Emery, Ann. Rev. Nucl. Sci. **22**, 165 (1972).
[3] T. Ohtsuki, K. Ohno, T. Morisato *et al.*, Phys. Rev. Lett. **98**, 252501 (2007).
[4] F. Raiola, T. Spillane, B. Limata *et al.*, Eur. Phys. J. A **32**, 51 (2007).
[5] W. Bambynek, H. Behrens, M. H. Chen *et al.*, Rev. Mod. Phys. **49**, 77 (1977).
[6] Y. A. Litvinov and F. Bosch, Rep. Prog. Phys. **74**, 016301 (2011).
[7] F. Bosch, Y. A. Litvinov and T. Stöhlker, Prog. Part. Nucl. Phys. **73**, 84 (2013).
[8] D. Atanasov, K. Blaum, F. Bosch *et al.*, J. Phys. B: At. Mol. Opt. **48**, 144024 (2015).
[9] M. Jung, F. Bosch, K. Beckert *et al.*, Phys. Rev. Lett. **69**, 2164 (1992).
[10] F. Bosch, T. Faestermann, J. Friese *et al.*, Phys. Rev. Lett. **77**, 5190 (1996).
[11] P. Walker and G. Dracoulis, Nature **399**, 35 (1999).
[12] P. M. Walker and G. D. Dracoulis, Hyperfine Interact. **135**, 83 (2001).
[13] G. D. Dracoulis, Phys. Scripta **T152**, 014015 (2013).
[14] Y. A. Litvinov, F. Bosch, H. Geissel *et al.*, Phys. Rev. Lett. **99**, 262501 (2007).
[15] N. Winckler, H. Geissel, Y. Litvinov *et al.*, Phys. Lett. B **679**, 36 (2009).
[16] D. R. Atanasov, N. Winckler, D. Balabanski *et al.*, Eur. Phys. J. A **48**, 22 (2012).
[17] Z. Patyk, J. Kurcewicz, F. Bosch *et al.*, Phys. Rev. C **77**, 014306 (2008).
[18] K. Siegień-Iwaniuk, N. Winckler, F. Bosch *et al.*, Phys. Rev. C **84**, 014301 (2011).
[19] F. Attallah, M. Aiche, J. F. Chemin *et al.*, Phys. Rev. Lett. **75**, 1715 (1995).
[20] W. R. Phillips, I. Ahmad, D. W. Banes *et al.*, Phys. Rev. Lett. **62**, 1025 (1989).
[21] W. R. Phillips, J. Copnell, D. W. Banes *et al.*, Phys. Rev. A **47**, 3682 (1993).
[22] A. Akber, M. W. Reed, P. M. Walker *et al.*, Phys. Rev. C **91**, 031301 (2015).
[23] E. M. Burbidge, G. R. Burbidge, W. A. Fowler *et al.*, Rev. Mod. Phys. **29**, 547 (1957).
[24] Yu.N. Novikov, H. Schatz, P. Dendooven *et al.*, Eur. Phys. J. A **11**, 257 (2001).
[25] B. A. Watson, T. T. Bardin, J. A. Becker *et al.*, Phys. Rev. Lett. **35**, 1333 (1975).
[26] J. Schirmer, D. Habs, R. Kroth *et al.*, Phys. Rev. Lett. **53**, 1897 (1984).
[27] J. Kramp, D. Habs, R. Kroth *et al.*, Nucl. Phys. **A474**, 412 (1987).
[28] C. Walz, H. Scheit, N. Pietralla *et al.*, Nature **526**, 406 (2015).
[29] M. Huyse, A. N. Andreyev, P. Van Duppen *et al.*, Nature **405**, 430 (2000).
[30] P. Dendooven, P. Decrock, M. Huyse *et al.*, Phys. Lett. B **226**, 27 (1989).
[31] D. R. Plante, W. R. Johnson and J. Sapirstein, Phys. Rev. A **49**, 3519 (1994).
[32] W. Johnson and G. Soff, Atom. Data Nucl. Data **33**, 405 (1985).
[33] Y. Litvinov, H. Geissel, Y. Novikov *et al.*, Nucl. Phys. **A734**, 473 (2004).
[34] Y. Litvinov, H. Geissel, T. Radon *et al.*, Nucl. Phys. **A756**, 3 (2005).

[35] L. Chen, Y. A. Litvinov, W. R. Plaß *et al.*, Phys. Rev. Lett. **102**, 122503 (2009).
[36] M. W. Reed, I. J. Cullen, P. M. Walker *et al.*, Phys. Rev. Lett. **105**, 172501 (2010).
[37] M. W. Reed, P. M. Walker, I. J. Cullen *et al.*, Phys. Rev. C **86**, 054321 (2012).
[38] M. W. Reed, P. M. Walker, I. J. Cullen *et al.*, J. Phys. Conf. Ser. **381**, 2058 (2012).
[39] L. Chen, P. M. Walker, H. Geissel *et al.*, Phys. Rev. Lett. **110**, 122502 (2013).
[40] F. Bosch, H. Geissel, Y. Litvinov *et al.*, Int. J. Mass. Spectrom. **251**, 212 (2006).
[41] L. Chen, W. R. Plass, H. Geissel *et al.*, Phys. Lett. B **691**, 234 (2010).
[42] F. Nolden, P. Hülsmann, Y. Litvinov *et al.*, Nucl. Instrum. Meth. A **659**, 69 (2011).
[43] M. S. Sanjari, P. Hüllsmann, F. Nolden *et al.*, Phys. Scripta **T156**, 014088 (2013).
[44] M. S. Sanjari, X. Chen, P. Hülsmann *et al.*, Phys. Scripta **T166**, 014060 (2015).
[45] B. Franzke, H. Geissel and G. Münzenberg, Mass Spectrom. Rev. **27**, 428 (2008).
[46] Y. Litvinov, F. Bosch, N. Winckler *et al.*, Phys. Lett. B **664**, 162 (2008).
[47] P. Kienle, F. Bosch, P. Bühler *et al.*, Phys. Lett. B **726**, 638 (2013).
[48] M. Steck, P. Beller, K. Beckert *et al.*, Nucl. Instrum. Meth. A **532**, 357 (2004).
[49] F. Nolden, K. Beckert, P. Beller *et al.*, Nucl. Instrum. Meth. A **532**, 329 (2004).
[50] H. Geissel, Y. Litvinov, F. Attallah *et al.*, Nucl. Phys. **A746**, 150c (2004).
[51] M. Hausmann, J. Stadlmann, F. Attallah *et al.*, Hyperfine Interact. **132**, 289 (2001).
[52] H. Geissel, Y. A. Litvinov, K. Beckert *et al.*, Eur. Phys. J.-Spec. Top. **150**, 109 (2007).
[53] H. Geissel, P. Armbruster, K. Behr *et al.*, Nucl. Instrum. Meth. B **70**, 286 (1992).
[54] B. Franzke, Nucl. Instrum. Meth. B **24/25**, 18 (1987).
[55] A. Musumarra, F. Farinon, C. Nociforo *et al.*, AIP Conf. Proc. **1165**, 415 (2009).
[56] C. Nociforo, F. Farinon, A. Musumarra *et al.*, Phys. Scripta **T150**, 014028 (2012).
[57] F. Herfurth, Z. Andelkovic, W. Barth *et al.*, Phys. Scripta **T166**, 014065 (2015).
[58] P. M. Hillenbrand, S. Hagmann, A. B. Voitkiv *et al.*, Phys. Rev. A **90**, 042713 (2014).
[59] P. M. Hillenbrand, S. Hagmann, D. H. Jakubassa-Amundsen *et al.*, Phys. Rev. A **91**, 022705 (2015).
[60] P. M. Hillenbrand, S. Hagmann, K. E. Stiebing *et al.*, Phys. Scripta **T166**, 014026 (2015).
[61] FAIR, FAIR Baseline Technical Report 1, GSI, Darmstadt (2006)., `http://www.fair-center.de/fileadmin/fair/publications_FAIR/FAIR_BTR_1.pdf`
[62] H. Geissel, H. Weick, M. Winkler *et al.*, Nucl. Instrum. Meth. B **204**, 71 (2003).
[63] A. Dolinskii, D. Berkaev, U. Blell *et al.*, Phys. Scripta **T166**, 014040 (2015).
[64] T. Stöhlker, Y. A. Litvinov and for the SPARC Collaboration, Phys. Scripta **T166**, 014025 (2015).
[65] T. Stöhlker, V. Bagnoud, K. Blaum *et al.*, Nucl. Instrum. Meth. B **365**, 680 (2015).
[66] J. Yang, J. Xia, G. Xiao *et al.*, Nucl. Instrum. Meth. B **317**, 263 (2013).
[67] A. Ozawa, T. Uesaka, M. Wakasugi *et al.*, Prog. Theor. Exp. Phys. **2012**, 03C009 (2012).
[68] T. Yamaguchi, Y. Yamaguchi and A. Ozawa, Int. J. Mass. Spectrom. **349-350**, 240 (2013).
[69] Y. Litvinov, S. Bishop, K. Blaum *et al.*, Nucl. Instrum. Meth. B **317**, 603 (2013).

EPJ Web of Conferences **123**, 04004 (2016) DOI: 10.1051/epjconf/201612304004
Heavy Ion Accelerator Symposium 2015

Nuclear lifetime measurements from data with independently varying observation times

T. J. Gray[1], M. W. Reed[1,a], G. J. Lane[1], A. Akber[1], Yu. A. Litvinov[2,3], and P. M. Walker[4]

[1]*Department of Nuclear Physics, Research School of Physics and Engineering, The Australian National University, Canberra, ACT 2601, Australia*
[2]*Max-Planck-Institut für Kernphysik, Saupfercheckweg 1, 69117 Heidelberg, Germany*
[3]*GSI Helmholtzzentrum für Schwerionenforschung, Planckstraße 1, 64291 Darmstadt, Germany*
[4]*Department of Physics, University of Surrey, Guildford, Surrey GU2 7XH, United Kingdom*

Abstract. A method is presented for obtaining ground-state and isomeric lifetimes from storage-ring data. The method published by Schmidt *et al.* is extended to the problems and solutions associated with applying their method to storage-ring data. These developed procedures are applied to real experimental data for ^{193}Re, and a mean-lifetime value of 37^{+9}_{-7} s is obtained. This is consistent with a previous measurement.

1 Introduction

As experiments start probing nuclei further away from the line of stability there is an inevitable drop in production cross-sections [1]. Therefore, obtaining estimates of the mean lifetimes of nuclei or nuclear states where poor or few measurements are available is becoming an important problem to address. In this paper, methods of lifetime measurement are investigated for use with storage-ring data. Different scenarios are explored through Monte-Carlo methods and tested on real data; the case of ^{193}Re is discussed.

This paper will focus on a particular application to storage-ring data, although in principle it can be applied to any experiment where the time of a single event can be measured. The data for ^{193}Re was produced using in-flight separated projectile fragmentation of a ^{197}Au beam on a ^{9}Be target. For details of the experiment, see Ref. [2]. The reaction products were injected into the Experimental Storage Ring (ESR) [3] and measured over many revolutions; the techniques used for these measurements are not discussed here but can be found in Refs. [4, 5].

A method for measuring lifetimes that has been used successfully in the past and been applied to storage-ring data [2, 6] is that presented by Schmidt *et al.* [7]. In Ref. [7], the mean lifetime is derived through what we have termed 'the unrestricted mean-lifetime method'. Here, each nucleus or nuclear state is measured from some start time to the point at which it is seen to decay, either by direct measurement, i.e. observation of a γ-ray, β or α particle, or inferred, as in the case of the storage-ring where a change in ion-orbital frequency implies a decay has occurred. In the case of storage-ring data, multiple ion events can be used, as long as each individual ion can be observed until it decays, and the individual times of survival measured [8]. A number (n_d) of these decays are measured and the observed 'survival' times summed together to give a total observation time (t_d); thus, simply dividing the two numbers gives the mean lifetime. The error on such a measurement is explained in Ref. [7] and it is implicit that this method can only be used when the observation time of each event is unrestricted, i.e. the each ion is observed until it decays.

However, in many situations with storage-ring data, an unrestricted observation time cannot be obtained, therefore an alternative method needs to be used to evaluate the lifetime. One such method using decay curves has been presented in Akber *et al.* [9]. However, this has limited applicability when only a few ions are observed as the decay curve is not well defined. Another method for extracting information from these data is suggested below.

2 Restricted Mean-Lifetime Method

In some experiments a fixed time for the observation will be used, for example, a 60-s measurement time. The unrestricted mean-lifetime method can be adjusted such that some information can be gained from the lifetime. As with the unrestricted version, the estimate of τ is obtained by dividing the total observation time by the number of decays seen. A Monte Carlo simulation was used to investigate the effectiveness of this method. Multiple simulations have been performed, but for the purposes of this paper τ is fixed at 60.4 s, and measurements are taken every 3.5 s; these numbers are chosen to approximate the conditions found in the data for ^{193}Re. A measurement corresponds to checking if the ion is still observed in the relevant time bin

[a]e-mail: matthew.reed@anu.edu.au

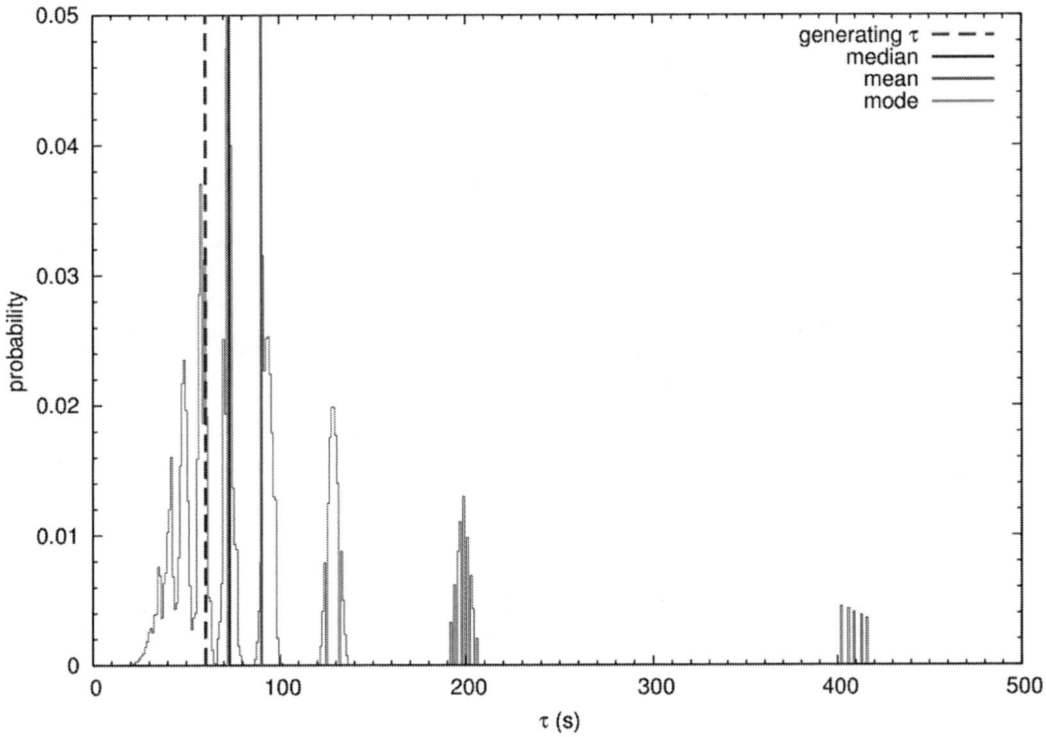

Figure 1. The probability density function obtained from the Monte Carlo simulation using $\tau = 60.4$ s, $n = 20$, and $b = 6$ (for a total observation time of 6×3.5 s $= 21.0$ s). The number of counts reflects the likelihood of obtaining the given τ from the average time method. The purple/dashed line shows the generating τ used in the simulation. The green line shows the maximum (mode) of the distribution, whilst the blue line indicates the median. The mean is marked by the magenta line.

– i.e. has not decayed. A number of different parameters can be varied, such as the initial population of ions n, and the number of possible time bins b (the largest possible observation time for one ion is thus $t_b \times b$ s, where t_b is the size of the time bin). The probability of decay in one time bin is $1 - \exp(-t_b/\tau)$, and a random number is generated to decide whether the ion has decayed in that time bin. This is repeated b times, and a time t is incremented by 3.5 s at each iteration where no decay is observed. If a decay is observed, the current t is added to a cumulative total t_{obs}, and the number of observed decays n_d is increased by 1. If no decay is observed, then $t_b \times b$ is added to the total t_{obs}. This process is repeated n times, and the final lifetime estimate is taken as

$$\overline{\tau} = \frac{t_{obs}}{n_d}.$$

By repeating this entire calculation numerous times, a probability density function can be found by histogramming the obtained mean-lifetimes ($\overline{\tau}$), as shown in Fig. 1. This represents the relative likelihood of obtaining an estimate of τ (along the x-axis) from the previously outlined method. Note the oscillatory behaviour in Fig. 1; each individual peak corresponds to a particular number of decays (since many ions survive for the full observation time). If n_d is low, the values of τ will be high, but as n_d becomes higher, the peaks overlap and form a smooth curve. The magnitude of each individual peak corresponds to the likelihood of that particular n_d occurring.

Once a histogram has been generated, an important question is which value best describes the outcomes of sampling from the distribution. For example, in the case of a gaussian curve, the mean is chosen as both the expected value, most likely outcome, and median. In less-ideal cases the decision may be unclear. The mean may be the most obvious choice as the expected value, however in some cases the mean is in fact impossible to obtain as an outcome. The mode, being the most likely outcome is also a possible choice, however cases where a maximum in the broad function corresponds to a minimum in the oscillations can result in a mode that either does not exist or is far away from the next most likely peak. For the purposes of this paper the median is chosen, as 50% of estimates will be above the median, and 50% below. See Fig. 1 for a comparison of the different averages.

As b and n increase, the probability density tends to a continuous distribution, and the difference between the actual τ and the median decreases, leading to estimates that are in general closer to the true τ. To account for the difference between the measured τ and the median of the dis-

DOI: 10.1051/epjconf/201612304004

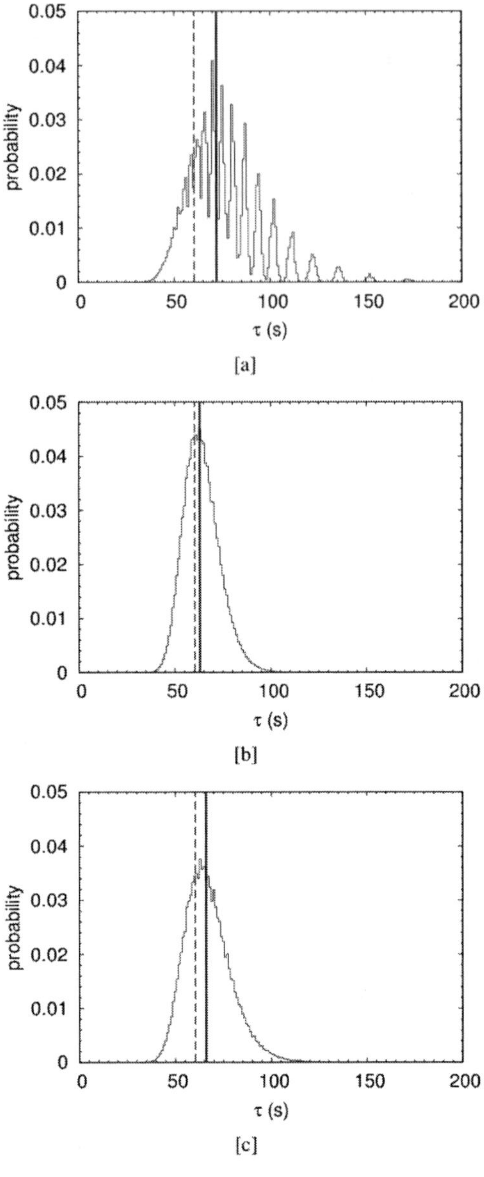

Figure 2. Panel (a) shows the histogram for the conditions $b = 6$, $n = 70$. The oscillations and skew can be clearly seen, with a significant difference between the median (blue) and generating τ (purple/dashed). Panel (b) shows the probability distribution produced by using $b = 21$, $n = 70$. In this case the method is clearly a much better estimate, with a smooth histogram and a median close to the generating τ. Panel (c) is produced using $b = 6$ for 35 ions, and $b = 21$ for the remaining 35. Qualitatively, this is a combination of (*a*) and (*b*), the distribution is much smoother than (*a*), but with a higher variance and skew than (*b*).

tribution a method has been developed for correcting the initial average time estimate to recover a best fit lifetime (Section 4).

There is uncertainty in the initial estimate obtained from experimental observations. The lifetime of each ion is not measured precisely due to the bin size. It is unknown where in the relevant bin the ion decays, and so each time measurement has an error of 3.5 s (the ^{193}Re case is discussed here). Thus the (experimental) error of $\overline{\tau}$ is given by

$$\Delta\overline{\tau} = \frac{\sqrt{3.5^2 + 3.5^2 + ... + 3.5^2}}{n_d} = \frac{3.5\sqrt{n}}{n_d}.$$

It is important to distinguish this experimental error from the statistical error inherent in the method itself. The first is a function of experimental limitations (namely precision of measurement), whilst the second is a result of the probabilistic nature of decay. This statistical error will be examined in Section 4.

3 Independently Varying Observation Time

A more common and widely applicable scenario is that the observation time is not fixed, but has a varying time window. The reason for this varying time window is the nature of the experiment itself. Generally, when an experiment is run, a region of interest can be monitored online, and then measurement times changed such that certain ions of interest can be observed for long time periods. This has a knock-on effect that the regions that are not monitored will also have the new measurement time. It is then a possibility that ions not in the observation region will be trapped for this extended period and more data taken. It is clear that these data can be reduced to the previous (fixed) case by cutting each observation off to the shortest observation time. However, this removes information that could otherwise have been used; what is proposed here is to use this additional information to obtain a more precise result.

Similar to before, a Monte Carlo simulation can be run, this time varying the observation times in a pre-defined configuration. For example, for a total population of 20 ions, the simulation can run the first 10 with $b = 5$, the next 5 with $b = 9$, 3 with $b = 15$, and the final 2 with $b = 30$. This has an effect of smoothing the probability density function such that it starts to tend to a continuous distribution. Figure 2c shows an example of this. Whilst this situation is not ideal, it is preferred to the fixed observation window. Although having a larger number of observation bins, larger observation window or more ions makes the most marked improvement, it is clear that the addition of a varying window length substantially cleans the probability distribution. However, in all cases a correction will need to be performed to obtain a correct estimate of the lifetime. This is discussed in Section 4.

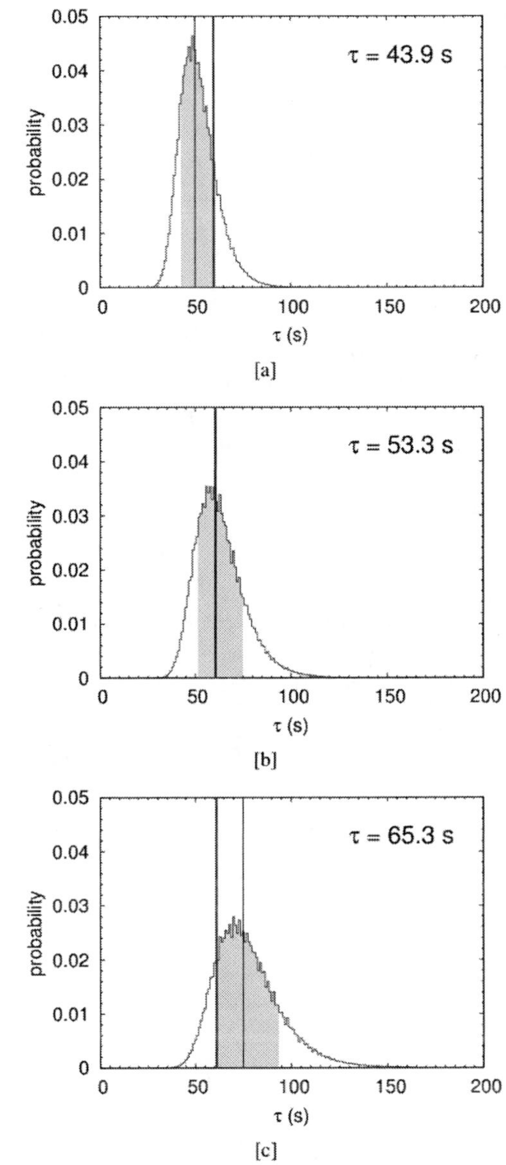

Figure 3. Probability distributions used to evaluate the uncertainties on the corrected value. The green line shows the initial uncorrected lifetime (and the blue line shows the median of the distribution, the shaded area is 68% of the distribution (1σ either side of the median). The number shown in each panel is the generating τ used to produce the histogram. Panel (a) shows the lower uncertainty estimate of the corrected τ. The lower uncertainty estimate is $\tau = 43.9$ s, i.e. the value used to generate the distribution. Panel (b) shows the estimate for the corrected τ. Panel (c) shows the upper uncertainty estimate of the corrected τ. All distributions are generated with the observation time configuration of the ^{193}Re data.

4 Correcting the Initial Mean-Lifetime Estimate

A methodology has been produced to correct for the inherent inaccuracy of the scenarios discussed above. The real experimental case of ^{193}Re has been used to as an example of how this can be applied to gain a genuine result. In the case of ^{193}Re, 74 single ion events were measured, with observation times ranging between 21.0 s and 234.5 s [10]. To obtain a real value for τ (the mean-lifetime) an adjustment to the initial mean-lifetime estimate must be made. To achieve this, a recursive Monte Carlo simulation is run. The initial estimate of $\tau = \tau_0 = \frac{t_{obs}}{n_d}$ is used to generate a histogram similar to Fig. 2 (c). The median of the distribution (m_0) is then compared with τ_0, and an offset $a_1 = m_0 - \tau_0$ evaluated. The Monte Carlo simulation is then run again with the new lifetime estimate of $\tau = \tau_1 = \tau_0 - a_1$. The recursive steps are summarised below:

$$a_{i+1} = m_i - \tau_0,$$

$$\tau_{i+1} = \tau_i - a_i,$$

with each median calculated from the histogram generated by setting $\tau = \tau_i$. The observation times used for each simulation are the same as those in the experimental data. After a number of steps, the result is a τ which produces a histogram with a median at τ_0, the initial estimate (Fig. 3b). This τ is the corrected value for the mean-lifetime of the ion.

To find error estimates for the corrected τ, the highest and lowest τ values for which it is reasonable to obtain an initial mean-lifetime estimate of τ_0 are found. For each histogram generated, define $l_i < m_i$ and $u_i > m_i$ such that 34% of the distribution area (1σ) is between l_i and m_i, and 34% (1σ) between m_i and u_i. Note that for a particular probability distribution, if $l_i \leqslant \tau_0 \leqslant u_i$ then it is reasonable to obtain an initial mean-lifetime estimate of τ_0. Conversely, if $l_i > \tau_0$ or $u_i < \tau_0$, it is unlikely ($> 1\sigma$) that τ_0 is obtained. Thus the highest reasonable τ is that which generates a distribution such that $l_i = \tau_0$. Similarly, the lowest reasonable τ is that which gives $u_i = \tau_0$.

If there is experimental error in τ_0, the high and low bounds of τ_0 are τ_{0h} and τ_{0l}. Then the highest reasonable (corrected) τ gives a distribution with $l_i = \tau_{0h}$, and the lowest reasonable τ gives $u_i = \tau_{0l}$. Figures 3(a) and (c) show this.

To find the lifetimes that generate the appropriate distributions, two more recursive processes are run, one converging to $\tau_0 = l_i$, and one to $\tau_0 = u_i$. This is achieved by having $a_{i+1} = l_i - \tau_0$, or $a_{i+1} = u_i - \tau_0$.

This process has been used on the 74 single ion observations of ^{193}Re. The initial τ_0 estimate obtained (before correction) is 60.4 ± 0.7 s. After applying the correction developed in section 4, a laboratory frame lifetime of $\tau = 53^{+12}_{-10}$ s was deduced. This lifetime needs to be further corrected to account for relativistic effects corresponding to the ion velocity of $\frac{v}{c} = 0.714$. The final value obtained is 37^{+9}_{-7} s; this can be compared to the only other result we know of, an unpublished value in Ref. [11] of 51^{+8}_{-8} s. Both values agree within uncertainties.

DOI: 10.1051/epjconf/201612304004

5 Discussion

This method has clear advantages over others which may be used for similar data sets. In comparison to curve fitting procedures as seen in [9], the corrected mean-lifetime method allows for all measured data to be used, without requiring all ions to be observed for the same period of time. This is particularly advantageous when only small numbers of decays are observed. It is also more widely applicable than methods which require an unrestricted lifetime. However, the limitations of the method are clear considering Figs. 1 and 2 (a). The discrete nature of the probability distributions for low n and b mean that any measurement obtained is inherently unreliable. Whilst a rough estimate of τ may be obtained, the applicability of the recursive correction outlined in Section 4 is less clear when the histogram is not smooth. In these cases it is not clear that the median is the correct average to use, however, the mode is equally problematic if a minimum in the fine oscillations coincides with the maximum of the broad function. The smoothing effects are difficult to quantify due to the large number of parameters, in particular, the wide variety of observation time configurations. However, on a case-by-case basis, the applicability of this method can be evaluated by modelling the observation time configuration observed in the experiment. An analytical solution to the problem is a subject for further study and beyond the scope of this paper.

6 Conclusion

The unrestricted mean-lifetime method has been adjusted to both restricted observation times, and independently varying observation times. The method allows all measured data to be used, as opposed to other methods which require a fixed observation time. A recursive algorithm corrects for the bias in the methods, and can be applied on an individual basis to specific experimental observation time configurations. The results obtained from this method for the ^{193}Re data are consistent with those obtained previously.

Acknowledgements

M.W.R. and G.J.L. were supported by grants from the Australian Research Council (DP120101417, DP140102986, and FT100100991). T.J.G. is grateful for the support from the RSPE Summer Scholarship Program. Y.A.L. is supported by the Helmholtz-CAS Joint Research Group (HCJRG-108). P.M.W. is supported by the UK-Science and Technology Facilities Council.

References

[1] J. Benlliure *et al.*, Nucl. Phys. A **660**, 87 (1999).
[2] M. W. Reed *et al.* Phys. Rev. C **86**, 054321 (2012).
[3] B. Franzke, Nucl. Instr. Meth. B **24/25**, 18 (1987).
[4] T. Radon *et al.*, Nucl. Phys. A **677**, 75 (2000).
[5] F. Bosch, J. Phys. B **36**, 585 (2003).
[6] M. W. Reed *et al.*, Phys. Rev. Lett. **105**, 172501 (2010).
[7] K. H. Schmidt *et al.*, Z. Phys. A **316**, 19 (1984).
[8] P. Kienle *et al.*, Phys. Lett. B **726**, 638 (2013).
[9] A. Akber *et al.*, Phys. Rev. C **91**, 031301 (2015).
[10] M. W. Reed *et al.*, unpublished.
[11] N. Alkhomashi, University of Surrey, Ph.D. thesis (2010).

EPJ Web of Conferences **123**, 04005 (2016) DOI: 10.1051/epjconf/201612304005
Heavy Ion Accelerator Symposium 2015

SPectrometer for Internal Conversion Electrons (SPICE) at TRIUMF-ISAC

J. Smallcombe[1,a], M. Moukaddam[1], L. J. Evitts[1,4], A. B. Garnsworthy[1], S. Hallam[1,4], C. Andreoiu[3], G. C. Ball[1], C. Bolton[1], R. Caballero-Folch[1], M. Constable[1], D. S. Cross[3], P. E. Garrett[2], G. Hackman[1], J. Henderson[1], R. Henderson[1], S. Ketelhut[1], R. Kruecken[1], L. Kurchaninov[1], J. Park[1,6], J. L. Pore[3], E. T. Rand[2], P. Ruotsalainen[1], J. K. Smith[1], C. E. Svensson[2], and M. Williams[1,5]

[1] *TRIUMF, 4004 Wesbrook Mall, Vancouver, B.C., V6T 2A3, Canada*
[2] *Department of Physics, University of Guelph, Guelph, Ontario, Canada, N1G 2W1*
[3] *Department of Chemistry, Simon Fraser University, Burnaby, British Columbia, Canada, V5A 1S6*
[4] *Department of Physics, University of Surrey, Guildford, Surrey, United Kingdom, GU2 7XH*
[5] *Department of Physics, University of York, Heslington, York, United Kingdom, YO10 5DD*
[6] *Department of Physics and Astronomy, University of British Columbia, Vancouver, British Columbia, Canada, V6T 1Z1*

Abstract. A new ancillary detector, SPICE (SPectrometer for Internal Conversion Electrons) has been constructed and recently commissioned for use with radioactive ion beams at the TRIUMF-ISAC II facility. SPICE is designed to be operated in conjunction with the TIGRESS High-Purity Germanium (HPGe) spectrometer to perform combined in-beam γ-ray and internal-conversion-electron spectroscopy. The main feature of SPICE is high efficiency over a wide range of electron energies from 100 to 3500 keV, with an effective reduction of beam-induced backgrounds. SPICE will be a powerful tool to measure conversion coefficients and $E0$ transitions in atomic nuclei. A recent in-beam commissioning experiment demonstrates the effectiveness of the basic design concept of SPICE in background suppression.

1 Introduction

In spherical nuclei, near closed shells, many excitations can be well explained by single-particle excitations in the microscopic picture of the nuclear shell model [1]. When attempting to model deformed nuclei, one must often rely on macroscopic collective models to explain the observed lowest energy excitations [2]. Typically the transition of nuclear structure between these two regimes occurs gradually with the addition of many valance nucleons, but in some special cases this transition is observed to occur with the addition of just a couple of pairs of valence nucleons. A key area of research is to understand the microscopic mechanisms responsible for the onset of deformation, and to understand the interplay between collective and single-particle excitations in deformed nuclei. This is particularly true when trying to explain the structure of nuclei identified as displaying shape coexistence at low excitation energy [3].

Low-lying 0^+ states with different deformation to the ground state are a primary feature of shape coexistence. Hence, understanding the nature of low-lying 0^+ states in deformed nuclei, and the excitations built upon them, is of particular interest. This can be experimentally challenging for several reasons: firstly the branching ratio for intra-band $E2$ transitions in bands built upon such excited 0^+ states are much smaller than the inter-band transitions de-exciting to the ground-state band, simply because of the larger transition energy (even when the intra-band $B(E2)$ value is much larger), secondly all the experimental observables, such as energy level spacing, branching ratios and reduced transition strengths, can be modified by band mixing with other configurations. This frequently obscures the true nature of 0^+ band-head states and confuses the assignment of structures built upon them.

SPICE (SPectrometer for Internal Conversion Electrons), at TRIUMF-ISAC, is a recently commissioned device used to perform in-beam electron spectroscopy with stable and radioactive ion beams. The spectrometer will be used to measure electric monopole transitions in exotic nuclei. Electric monopole transition strengths, $\rho^2(E0)$, are crucial observables in the interpretation of nuclear structure. The electric monopole strength between two states of the same spin and parity is directly proportional to the difference in their mean-squared charge radii and the degree of mixing between the two configurations. SPICE will enable the identification of excited 0^+ states in exotic nuclei and the determination of $\rho^2(E0)$ values, which will help develop our understanding of shape coexistence away from the line of stability.

[a] e-mail: jsmallcombe@triumf.ca

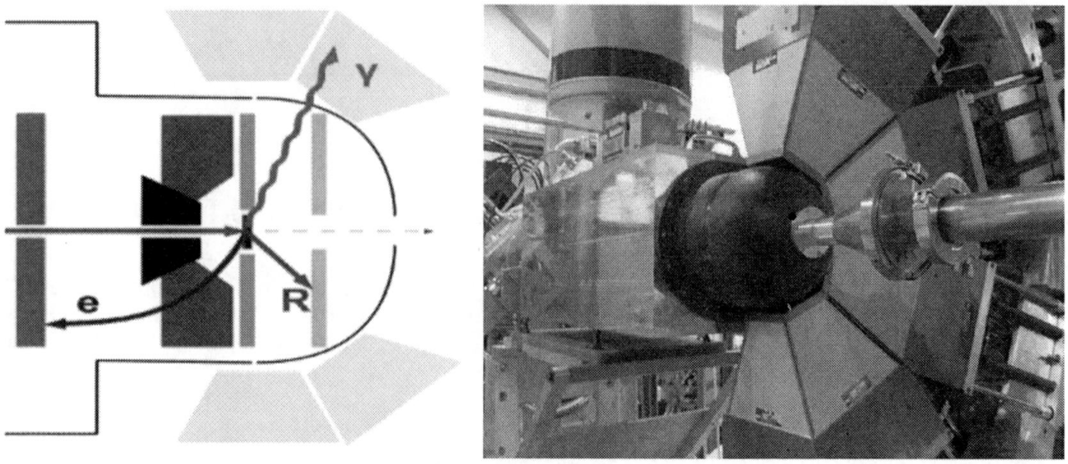

Figure 1. Left: Schematic of the primary components of SPICE as described in the text, Si(Li) detector (green), photon-shield (black), magnetic lens (red), recoil detector (yellow) and TIGRESS High-Purity Germanium (HPGe) clovers (blue). Lines show typical path of electrons (e), γ rays (γ), recoiling heavy ions (R) and radioactive beam (incident from left). **Right:** Photograph of the SPICE chamber mounted alongside one half of the TIGRESS array.

2 SPectrometer for Internal Conversion Electrons

SPICE consists of an electron detector, photon shield and a magnetic lens. A schematic of the arrangement is shown in Fig. 1. Electrons are detected in an annular lithium-drifted silicon (Si(Li)) detector of 6.1 mm thickess which is cooled by a liquid nitrogen coldfinger to −100°C. The considerable thickness of the detector allows for complete energy collection, hence increasing detection efficiency, even for highly penetrating electrons with several MeV of energy. Cooling allows the detector to achieve a resolution of FWHM < 5 keV; this fact is demonstrated in Fig. 2 which show the raw electron spectrum taken with a ^{207}Bi source. The detector has an outer diameter of 10 cm and a central hole of diameter 10 mm, through which radioactive beams delivered from ISAC pass before striking the reaction target. The detector is segmented into 120 individual segments arranged as 12 azimuthal sectors and 10 rings. The electron detector is located 111.3 mm upstream of the target and is shielded from direct view of the target by a multi-layer (tungsten-tantalum alloy, tin, copper) photon shield, to block direct irradiation by X-rays and γ rays. The magnetic lens collects and directs electrons emitted from the target position around the photon shield to the electron detector. The magnetic lens consists of four sub-assemblies of high grade neodymium iron boron (NdFeB) rare-earth permanent magnet, arranged around the photon shield. The lens is oriented such that the high-Z material has a minimal effect on γ-ray detection efficiency of the coupled TIGRESS array [5], as each sub-assembly is aligned with the suppression shield of the HPGe Clover detectors. Through a detailed design process [4] three magnetic lens designs have been developed (and two built) to cover the energy range from 100 keV to 4 MeV. An absolute efficiency greater than 15% is achieved, which far exceeds the natural solid angle coverage of the detector. A variety of particle detectors can be coupled with SPICE in the downstream hemisphere of the target chamber to detect coincident heavy ions. An S3 type silicon CD detector has been used so far.

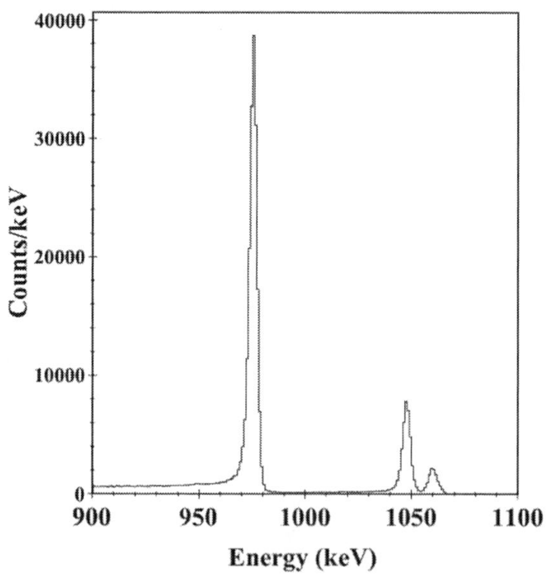

Figure 2. Raw electron spectrum in SPICE from a ^{207}Bi source, showing three distinct peaks of the K, L and M electrons associated with the 1063.7-keV γ ray. This clearly demonstrates the impressive optimum experimental resolution SPICE can achieve with FWHM = 4.5 keV at 1 MeV.

DOI: 10.1051/epjconf/201612304005

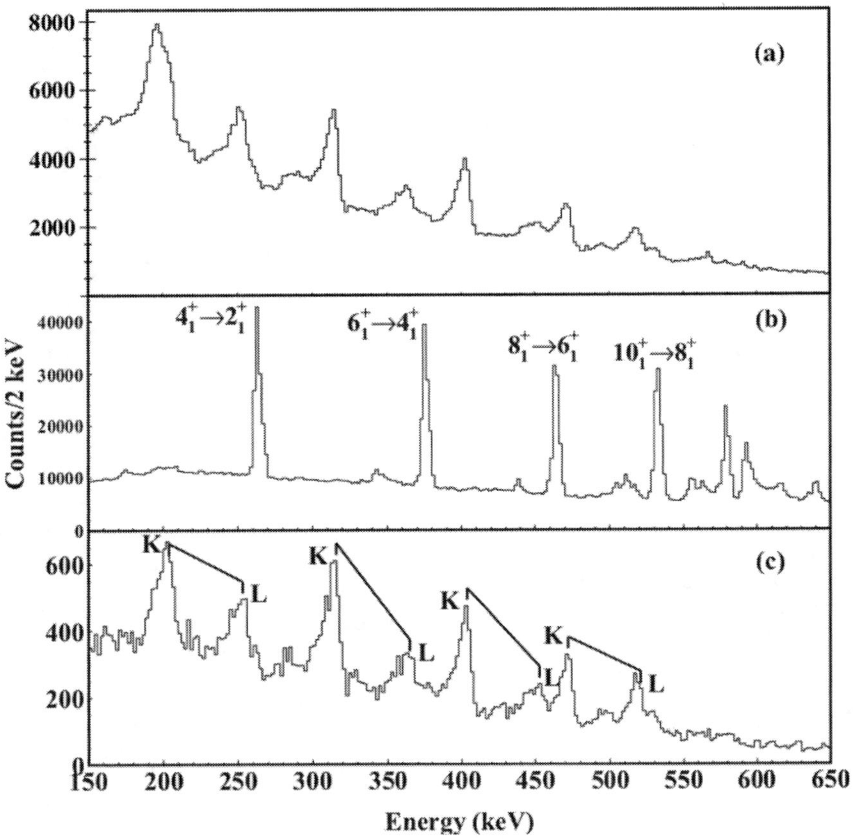

Figure 3. In-beam data from 20 hours of data for a 5.7-MeV/u ^{12}C beam at 1×10^8 pps, impinging on a 4-mg/cm^2 ^{152}Sm target. The raw electron singles spectrum from SPICE (a) is dominated by transitions from the yrast band of ^{160}Er following fusion evaporation reactions. This raw singles spectrum shows that even with no coincidence requirements, SPICE still yields a formidable peak-to-total ratio for a conversion electron spectrometer. The gated γ-ray (b) and internal conversion electron (c) spectra are produced by applying a coincidence requirement with an intra-band γ-ray, within the yrast band of ^{160}Er, on γ-γ and electron-γ matrices. The intra-band γ rays, *K* electrons (binding energy 58 keV) and *L* electrons (binging energy 10 keV) of ^{160}Er are labeled. Peak ratios between these two spectra, combined with the detector efficiencies, can be used to produce conversion coefficient measurements.

3 Commissioning Experiment

A beam of ^{12}C at an energy of 5.7 MeV/u and an intensity of 10-100 ppA (about 1×10^8 – 1×10^9 pps) was directed at targets of ^{152}Sm (4 mg/cm^2) and ^{196}Pt (2.9 mg/cm^2). The beam energy was such that both Coulomb excitation and fusion-evaporation reactions were possible. This setup produced several data sets with different kinematics, multiplicities and target effects. The raw, un-gated, electron spectra and γ-ray spectra for each target were dominated by the fusion-evaporation channels $^{152}Sm(^{12}C,4n)^{160}Er^*$ and $^{196}Pt(^{12}C,4n)^{204}Po^*$, respectively. The platinum target was backed by a 10-mg/cm^2 stopper foil of natural Pb to catch recoiling nuclei, so that isomeric states in ^{204}Po could be studied. This will be discussed in a future publication. Figure 3a shows the fusion-evaporation dominated spectrum for $^{152}Sm+^{12}C$; even this raw spectrum, without coincidence gating, kinematic correction or background subtraction, shows reasonable peak-to-total ratios and contains sufficient distinct peaks for good fitting. Figures 3b and 3c demonstrate the effectiveness of γ-ray coincidence gating. The high efficiency and resolution of TIGRESS allows for clean gating of data and further the identification of and subtraction of background (not shown here). Requiring detection of a scattered ^{12}C beam ion in the downstream silicon detector, in coincidence with detected electrons and γ rays, cleanly selects the Coulomb excitation channels $^{152}Sm(^{12}C,^{12}C)^{152}Sm^*$ and $^{196}Pt(^{12}C,^{12}C)^{196}Pt^*$. Figure 4 shows the effectiveness of a coincident heavy ion gate for the ^{196}Pt target. These spectra are gated only on a recoiling ion in the downstream silicon detector. Without further subtraction, this is sufficient to remove all fusion-evaporation channels and show only the Coulmb excitation data. By combining both heavy-ion and γ-ray coincidence gating shown here, as well as background subtraction and kinematic corrections currently being implemented, SPICE will be an extremely clean, high-resolution and high-efficiency device for internal electron spectroscopy.

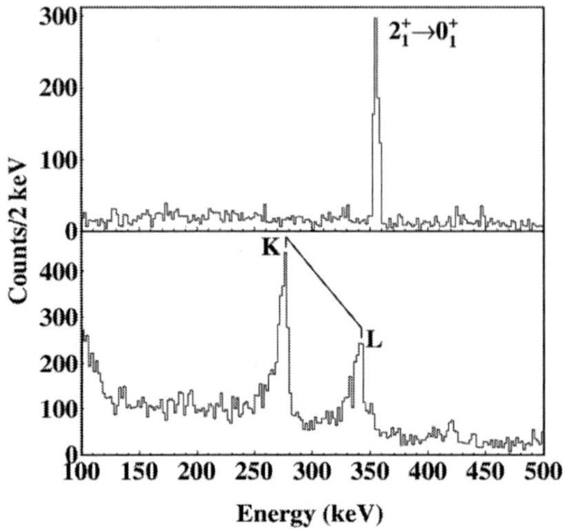

Figure 4. In-beam γ-ray (top) and internal conversion electron (bottom) data showing 2 hours of Coulomb excitation data for a 5.7-MeV/u ^{12}C beam at 1×10^8 pps, impinging on a 2.9-mg/cm^2 ^{196}Pt target, requiring a coincident heavy ion detected in the downstream recoil detector. The ^{196}Pt $2_1^+ \rightarrow 0_1^+$ *K*-electron and *L*-electron peaks are both prominent without background subtraction or γ-ray gating.

4 Outlook

The spectra presented here represent only ~1% of the commissioning experiment data. Analysis of the full dataset, to demonstrate the first *E*0 measurement, is currently underway. The effectiveness of SPICE will be further improved by ongoing development of analysis techniques, a following publication will demonstrate the effectiveness of kinematic correction and add-back in the SPICE detector. Development of alternative downstream detectors is in progress to study states of interest populated using a variety of reaction mechanisms. The first experimental measurements with SPICE and TIGRESS are expected to take place in 2016.

Acknowledgements

The SPICE collaboration would like to thank SEMIKON Detector GmbH, for developing the segmented Si(Li) detector. Construction of SPICE has been funded by the Canada Foundation for Innovation and the Ontario Ministry of Research and Innovation. This work was supported in part by Natural Sciences and Engineering Research Council (NSERC) of Canada. TRIUMF receives federal funding via a contribution agreement through the National Research Council of Canada.

References

[1] M. G. Mayer and J. H. D. Jensen, *Elementary Theory of Nuclear Structure* (Wiley, New York, 1955).
[2] S. Bohr and B. R. Mottelson, *Nuclear Structure II* (W. A. Benjamin, 1975).
[3] K. Heyde and J. L. Wood, Rev. Mod. Phys. **83**, 1467 (2011).
[4] S. Ketelhut, L. J. Evitts, A. B. Garnsworthy, *et al.*, Nuc. Inst. Meth. A **753**, 154 (2014).
[5] G. Hackman, C. E. Svensson, Hyperfine Interactions **225**, 241 (2014).

EPJ Web of Conferences **123**, 04006 (2016) DOI: 10.1051/epjconf/201612304006
Heavy Ion Accelerator Symposium 2015

Recent advances in β-decay spectroscopy at CARIBU

A. J. Mitchell[1,a], P. Copp[2], G. Savard[3,4], C. J. Lister[2], G. J. Lane[1], M. P. Carpenter[3], J. A. Clark[3], S. Zhu[3], A. D. Ayangeakaa[3], S. Bottoni[3], T. B. Brown[2], P. Chowdhury[2], T. W. Chillery[2,5], H. M. David[3,b], D. J. Hartley[6], E. Heckmaier[7], R. V. F. Janssens[3], K. Kolos[8], F. G. Kondev[9], T. Lauritsen[3], E. A. McCutchan[10], E. B. Norman[8,11], S. Padgett[8], N. D. Scielzo[8], D. Seweryniak[3], M. L. Smith[1], and G. L. Wilson[2]

[1] *Department of Nuclear Physics, The Australian National University, Canberra, ACT 2601, Australia*
[2] *Department of Physics and Applied Physics, University of Massachusetts Lowell, Lowell, Massachusetts 01854, USA*
[3] *Physics Division, Argonne National Laboratory, Argonne, Illinois 60439, USA*
[4] *Department of Physics, University of Chicago, Chicago, Illinois 60637, USA*
[5] *Department of Physics, University of Surrey, Guildford, GU2 7XH, United Kingdom*
[6] *Physics Department, The United States Naval Academy, Annapolis, Maryland 21402, USA*
[7] *Department of Physics and Astronomy, University of California, Irvine, California 92697, USA*
[8] *Lawrence Livermore National Laboratory, Livermore, California 94550, USA*
[9] *Nuclear Engineering Division, Argonne National Laboratory, Argonne, Illinois 60439, USA*
[10] *National Nuclear Data Center, Brookhaven National Laboratory, Upton, New York 11973, USA*
[11] *Department of Nuclear Engineering, University of California, Berkeley, California 94720, USA*

Abstract. β-decay spectroscopy of nuclei far from stability can provide powerful insight into a broad variety of topics in nuclear science, ranging from exotic nuclear structure phenomena, stellar nucleosynthesis processes, and applied topics such as quantifying "decay heat" discrepancies for advanced nuclear fuel cycles. Neutron-rich nuclei approaching the drip-line are difficult to access experimentally, leaving many key examples largely under studied. The CARIBU radioactive beam facility at Argonne National Laboratory exploits spontaneous fission of ^{252}Cf in production of such beams. The X-Array and SATURN decay station have been commissioned to perform detailed decay spectroscopy of low-energy CARIBU beams. An extended science campaign was started during 2015; with projects investigating nuclear shape changes, collective octupole vibrations, β-delayed neutron emission, and decay-scheme properties which could explain the reactor antineutrino puzzle. In this article we review the current status of the setup, update on the first results and recent hardware upgrades, and look forward to future possibilities.

1 Introduction

In recent years, significant effort in experimental nuclear physics has been expanded to access more and more exotic regions of the nuclear chart lying far from stability. The greatest hindrance to such experiments is the diminutive production cross sections associated with radioactive ion beams (RIBs). One method to circumvent this issue is to build highly-efficient experimental setups, capable of detecting as many of these rare events as possible.

Neutron-rich nuclei with $100 < A < 160$ present an interesting region with regard to a broad range of topics in nuclear-physics research. For example, ^{132}Sn is often discussed as an example of a doubly-magic nucleus [1]. Studies performed on its neighbouring nuclides can offer detailed insight into single-particle excitations and proton-neutron interactions with large core neutron excesses. Dramatic ground-state shape changes due to collective motion are known to occur in the neutron-rich $A \sim 100$ region [2]. Many of the most interesting cases happen to be refractory elements, which are notoriously difficult to access through conventional RIB-production methods. The path of the r-process is also predicted to lie within this area [3]. Nuclear ground-state and isomer masses and decay half-lives are central to r-process and elemental abundance calculations, with sensitivity studies highlighting the need for more accurate measurements to be performed (e.g. Ref. [4]). The region also incorporates products of the fission process, which is pertinent to the understanding of decay heat in advanced fuel cycle processes [5] and the so-called 'antineutrino anomaly' [6].

The most direct route to accessing neutron-rich RIBs is through the fission of heavy nuclei. The CAlifornium Rare Ion Breeder Upgrade (CARIBU) is a RIB facility located at Argonne National Laboratory [7]. At the heart of the CARIBU system is a ~1-Ci ^{252}Cf source. This nucleus primarily α decays with a half-life of approximately 2.6 years. However, it also possesses a spontaneous fission branch of about 3% which is exploited by CARIBU.

[a] e-mail: aj.mitchell@anu.edu.au

[b] Present address: GSI Helmholtzzentrum für Schwerionenforschung GmbH, D-64291 Darmstadt, Germany

Since ^{252}Cf possesses greater A and Z than other actinides typically used in ISOL-based facilities, such as proton- or neutron-induced fission of ^{235}U, CARIBU beams have the potential to access unique regions of the nuclear landscape. There is also the added benefit of greatly-enhanced refractory element extraction from the ^{252}Cf fission source.

Fission fragments extracted from the source are thermalised in a gas catcher that was developed at Argonne. Here, they are slowed in high-purity helium gas, and a combination of DC and RF fields focus them into a low-emittance beam. The species of interest is selected in the isobar separator, where it is either sent for charge breeding and acceleration through the ATLAS facility, or directed to the low-energy experimental area. A new β-decay spectroscopy station has been built for use with low-energy CARIBU beams. The first dedicated experimental campaign to be conducted with the new setup is the focus of this article.

2 First results

The decay station consists of two main components. The X-Array is a highly efficient array of five High-Purity Ge clover detectors mounted on an adjustable frame. The frame itself was designed to provide flexibility, enabling the clovers to be easily positioned or removed as necessary to accommodate auxiliary detectors. Large clover crystals located in close proximity to the collection point result in a high γ-ray detection efficiency of ~12% at 1 MeV. Utilising the 'add-back' capabilities of the clovers increases the measured peak-to-total ratio by reconstructing events that are scattered between crystals, thereby improving photopeak efficiency and reducing the Compton background. The second component is SATURN, a tape-transport system for removal of long-lived activity with plastic scintillators surrounding the collection point for detection of β particles. A full description of the decay station can be found in Ref. [8].

Commissioning of the decay station was completed in 2014 [9]. Much of the early work was conducted with beams of neutron-rich Cs isotopes. Chemistry of this region enables excellent m/q selection for Cs ions as they favour extraction in a 1^+ charge state from the gas catcher, whereas the neighbouring isobars favour a 2^+ charge state. Results from the ^{146}Cs β-decay experiments have recently been published [10].

3 Recent developments

The first extended low-energy decay campaign took place in 2015. One of the experiments we performed was focused on the $A = 100$ region which is well known for dramatic shape changes. The Sr and Zr isotopes undergo a dramatic change from vibrational to rotational character between $N = 58$ and $N = 60$. However, with just two protons more, the Mo isotopes are seen to undergo a more gradual shape transition. The heaviest Mo isotopes exhibit some of the lowest 2_2^+ energies measured, suggesting the existence of triaxial deformation, for which accurate γ- to ground-state-band branching ratios are desirable.

However, these nuclei are difficult to access experimentally due to their exotic and refractory nature. High-spin band structures have been observed from prompt-fission studies, whereas little is known about collective structures at low spin and excitation. β decay of Nb is an ideal probe of the low-energy structure of Mo. Detailed data sets of ^{104,106}Nb $\rightarrow$ ^{104,106}Mo β decay were taken for five days and are currently under analysis. A sample of the data for ^{104}Mo is presented in Fig. 1. The spectrum shows a projection of the $\beta - \gamma - \gamma$ coincidence matrix, gated on the 620-keV γ-ray transition connecting the $2_\gamma^+ \rightarrow 2_{gs}^+$ levels. A partial level scheme has been constructed using this information and is also shown. Spins and parities quoted here are taken from Ref. [11]. The ground state of ^{104}Nb is tentatively assigned $J^\pi = (1^+)$ in the literature. Strong population of the high-spin states that would be highly forbidden in β decay from the low-spin ground state is due to the occurrence of a β-decaying isomer in the parent which must have $J \approx 5$.

Each recorded event is timestamped relative to the beginning of a predetermined time cycle. The cycling used for the ^{104}Nb decay experiment consisted of 10-s beam-on/beam-off intervals to study the ground-state and isomer decay with half-lives of 0.94(4) s and 4.9(3) s, respectively [11]. The growth-and-decay curve for the 368-keV β-gated γ-ray transition ($4_{gs}^+ \rightarrow 2_{gs}^+$) is shown in Fig. 2. Since a $1^+ \rightarrow 4^+$ β decay is second forbidden, the 4^+ levels in ^{104}Mo must be predominantly populated by the isomer decay. By fitting an exponential decay plus linear background function to the 'decay' portion of the curve, a half life of 0.97(1) s was extracted for the isomer from these data. This is in excellent agreement with the adopted value of 0.94(4) [11].

During this campaign, auxiliary detectors were installed on the X-Array, in place of one of the clovers, for the first time. Five $LaBr_3$(Ce) detectors were installed for two of the experiments with the aim of measuring excited-state lifetimes shorter than 100 ps. An array of 16 Cs_2LiYCl_6:Ce (CLYC) scintillator detectors, built by the University of Massachusetts Lowell, was also commissioned during the campaign. Conventional CLYC scintillators perform as γ-ray spectrometers with good energy resolution, but also as thermal neutron detectors, via the ^{6}Li(n, α) reaction, as they offer excellent pulse-shape discrimination [12]. They have also been found to be sensitive to fast neutrons via the ^{35}Cl(n, p) reaction. In the Lowell array, the thermal neutron response is suppressed by depleting the ^{6}Li content [13]. This makes them ideal for studying fission neutrons and β-delayed neutron emission with neutron-rich CARIBU beams.

In total, seven experiments investigating a broad range of topics across nuclear-reactor physics, β-delayed neutron emission, nuclear octupole vibrations and shape coexistence were performed during the campaign, which spanned 28 days of low-energy beam time. Analyses of the data are ongoing, and we are excited about the prospect of results the campaign may yield in the near future.

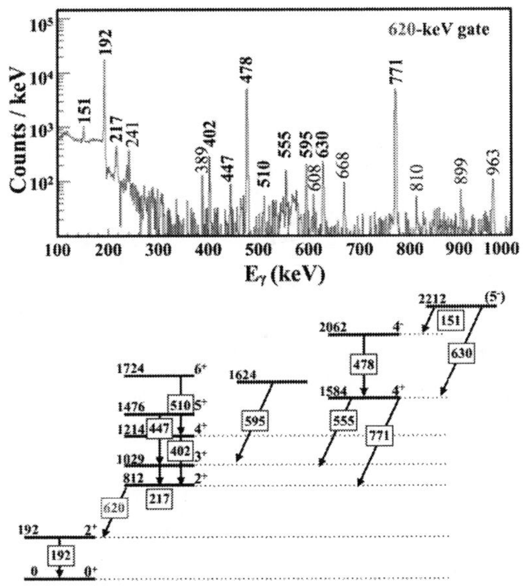

Figure 1. [Colour online] (Top) Projection of the $\beta - \gamma - \gamma$ coincidence matrix gated on the 620-keV γ-ray transition in ^{104}Mo. (Bottom) Partial level scheme (not to scale) constructed from the 620-keV γ-ray gate. The γ rays included in the level scheme are labelled in bold in the spectrum. The other labelled γ rays are weak branches that have not been included in this level scheme. Spins and parities of the levels included here are from Ref. [11].

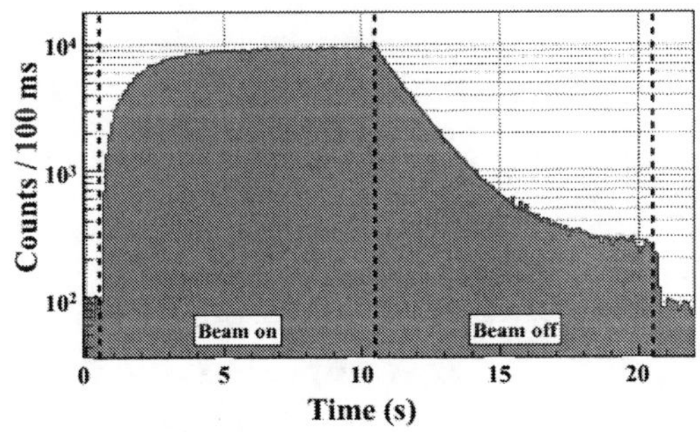

Figure 2. [Colour online] Example of a growth-and-decay curve for the time cycle described in the text. These data are for the 368-keV β-gated γ-ray transition in ^{104}Mo.

4 Outlook

A number of additional measurements have been performed since the initial campaign. The majority of these have been with beam species with the largest fission yields. Two issues for subsequent experiments that aim to study weaker beams are the isolation of the isotope of interest from the neighbouring isobars and the backgrounds induced by radiation from the CARIBU source. A new MR-TOF separator has been installed on the low-energy beamline. This enables much more effective mass separation which is crucial for studies of exotic isotopes. The next stage of experiments will involve relocating the low-energy experiments to the adjacent hall which used to house the tandem injector. This will be a significant undertaking, and will require construction of new beam lines and other infrastructure. The benefits after this work is completed should be significant, opening up possibilities to study nuclei pushing out closer to the neutron drip line.

Acknowledgements

The authors acknowledge the Physics Support Group at Argonne National Laboratory and engineering work of the Submillimeter-Wave Technology Laboratory, University of Massachusetts Lowell. This material is based upon work supported by the Australian Research Council Discovery Project 120104176 (ANU), the U.S. Department of Energy, Office of Science, Office of Nuclear Physics under Grants No. DE-FG02-94ER40848 (UML) and No. DE-AC02-98CH10886 (BNL), and Contracts No. DE-AC02-06CH11357 (ANL) and No. DE-AC52-07NA27344 with funding support from the Laboratory Directed Research and Development Program under project 14-LW-087 (LLNL). Also supported by the U.S. National Science Foundation under Grant No. PHY-1203100 (USNA), and the U.S. Department of Energy National Nuclear Security Administration under Grant No. DE-NA-0000979 through the Nuclear Science and Security Consortium (UCB). This research used resources of ANL's ATLAS facility, which is a DOE Office of Science User Facility.

References

[1] K. L. Jones *et al.*, Nature **465**, 454 (2010).
[2] E. Cheifetz *et al.*, Phys. Rev. Lett. **25**, 38 (1970).
[3] E. Margaret Burbidge *et al.*, Rev. Mod. Phys **29**, 547 (1957).
[4] M. Mumpower *et al.*, Phys. Rev. C. **85**, 045801 (2012).
[5] N. Hagura, T. Yoshida, and T. Tachibana, J. Nucl. Sci. Techn. **43**, 497 (2006).
[6] G. Mention *et al.*, Phys. Rev. D **83**, 073006 (2011).
[7] G. Savard *et al.*, Nucl. Instrum. Meth. B **266**, 4086 (2008).
[8] A. J. Mitchell *et al.*, Nucl. Instrum. Meth. A **763**, 232 (2014).
[9] A. J. Mitchell *et al.*, JPS Conf. Proc., **6**, 020013 (2015).
[10] A. J. Mitchell *et al.*, Phys. Rev. C. **93**, 014306 (2016).
[11] J. Blachot, Nucl. Data Sheets **108**, 2035 (2007).
[12] N. D'Olympia *et al.*, Nucl. Instrum. Meth. A **694**, 140 (2012).
[13] N. D'Olympia *et al.*, Nucl. Instrum. Meth. A **714**, 121 (2013).

DOI: 10.1051/epjconf/201612304007

Decay spectroscopy with Solenogam at the ANU Heavy Ion Accelerator Facility

M. S. M. Gerathy[1,a], M. W. Reed[1], G. J. Lane[1], T. Kibédi[1], S. S. Hota[1], and A. E. Stuchbery[1]

[1]*Department of Nuclear Physics, Research School of Physics and Engineering, The Australian National University, Canberra, ACT 2601, Australia*

Abstract. Solenogam is a recoil spectrometer designed and constructed for use at the Australian National University (ANU) Heavy-Ion Accelerator Facility (HIAF). The design enables the study of nuclear excitations populated by the decay of long-lived states such as isomers and radioactive ground states. Solenogam is comprised of high-sensitivity γ-ray and electron detector arrays coupled to a new 8-T solenoid. While the installation of the 8-T solenoid proceeds, off-line measurements have been made to characterise Solenogam's performance. Gamma-electron coincidences in the electron capture decay of ^{182}Re into ^{182}W were used to investigate conversion coefficients and γ-e$^-$ angular correlations. The measured conversion coefficients show good agreement with theoretical calculations and have been used to extract $E0/E2$ mixing ratios for a number of $J \rightarrow J$ transitions. The angular correlations measured by the array are in qualitative agreement with theoretical calculations. However, the magnitudes of the correlations are attenuated by approximately 40% for reasons unknown at present. These results are the first full use of the Solenogam system for γ-e$^-$ coincidence measurements and have proven that the system is capable of highly-sensitive internal conversion analysis of complex decays.

1 Introduction

Shape coexistence is a phenomenon observed in nuclei near closed proton and mid neutron shells (e.g $Z \sim 82$, $N \sim 104$). In these nuclei, particle/hole excitations can lead to competing nuclear shapes [1]. One characteristic of coexisting shapes is the presence of multiple rotational bands with different energy spacings dependent on the moment of inertia. Transitions between these rotational structures with $J \rightarrow J$ spin changes are important as they proceed via $E0$, $M1$ and $E2$ components, whose strengths depend strongly on the wavefunctions (and hence shapes) of the initial and final states. As these transitions can proceed with significant $E0$ character, electron spectroscopy is required to fully characterise these transitions.

The current spectrometers available at the ANU-HIAF are the Compton suppressed array (CAESAR) [2] and the Superconducting electron spectrometer (Super-e) [3]. However, neither are suitable for extracting $E0$ strengths from complex data sets. While the CAESAR array is capable of measuring γ-γ coincidences, it lacks the capability for simultaneous measurements of electrons. Conversely, although the Super-e spectrometer can perform e$^-$-e$^-$ coincidence measurements if the electrons are within the acceptance window of the magnet, it is not designed for γ-ray spectroscopy and has only limited e$^-$-γ coincidence capability. The new Solenogam array is capable of combining the benefits of both systems, providing simultaneous access to γ-γ, γ-e$^-$ and e$^-$-e$^-$ coincidences. The array consists of six, 2-mm Si(Li) electron detectors and up to seven γ-ray detectors (including four compton suppressed detectors). Using standard spherical coordinates, the detectors are located at $\phi = 0°$, $\pm 60°$, $\pm 120°$, and $180°$ around the beam axis. The Si(Li) detectors are located upstream of the focal point at $\theta = 135°$ to the beam axis while the γ-ray detectors are located downstream at $\theta = 45°$ (A schematic diagram is shown in Fig. 1).

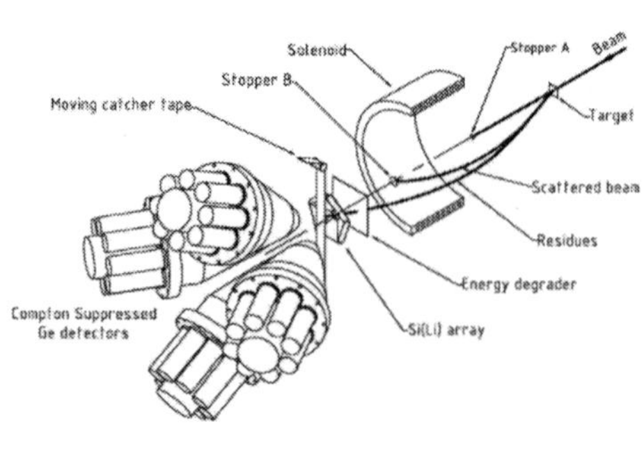

Figure 1. Schematic diagram of the Solenogam system. Note that the three unsuppressed HPGe detectors are not shown.

In the current work, two measurements were made with the array. The first used the Si(Li) array coupled with two high purity germanium (HPGe) gamma-ray detectors. This was subsequently increased to three HPGe and one low energy photon spectrometer (LEPS) detector

[a]e-mail: matthew.gerathy@anu.edu.au

 DOI: 10.1051/epjconf/201612304007

for the second measurement. It should be noted that the third HPGe detector had a non-standard placement in the array and was located at $\theta \approx 45°$ and $\phi = 90°$. In both measurements, one of the HPGe detectors was Compton suppressed. Energy resolution in matrices that sum across all detectors are $\approx$ 3 keV (0.2%) for the HPGe detectors and $\approx$ 4 keV (0.4%) for the Si(Li) detectors for the 1408-keV ^{152}Eu decay line and corresponding K conversion line, respectively.

Sources of ^{182}Re with activities of between 1 and 3 μCi were created for offline measurement by the irradiation of 1-mg/cm^2 and 2-mg/cm^2 ^{176}Yb foils. The reaction chosen for this was ^{176}Yb(^{11}B,5n)^{182}Re using 60-MeV ^{11}B ions. The 64-hour halflife of the ^{182}Re parent made offline measurements feasible and measurements were taken for approximately four halflives. During analysis it was found that the 2-mg/cm^2 target was not centred at the focus of the spectrometer, therefore, all spectroscopic data presented here is from the 1 mg/cm^2 source, unless otherwise stated.

2 Results

2.1 ^{182}W

The structure of ^{182}W has been studied extensively with a wide range of spectroscopic information available for comparison [4–11]. A number of rotational structures are populated in the ^{182}Re decay. These are built upon the ground state, and the β (0^+), γ (2^+), and octupole (3^-) vibrations at 0, 1135, 1221 and 1289 keV, respectively. The 4^- band head at 1553 keV is a multi-quasiparticle state [11]. Unfortunately, the excited 0^+ state at 1135 keV was not populated in the decay so pure $E0$ transitions were not produced. Gamma-rays were assigned to ^{182}W by comparison with previously measured energies and intensities [9]. Electron lines were assigned using the known tungsten K, L and M binding energies of 69.5, 11.5 and 2.2 keV, respectively [12]. The γ-ray spectrum showed minimal contamination and analysis of characteristic X-rays using the LEPS detector identified tungsten as the major product.

2.2 Spectra and conversion coefficients

The level scheme of ^{182}W is rather complex involving many low-energy transitions (100-400 keV), causing both ungated γ and electron spectra to be contaminated (see Fig. 2). Solenogam, however, is able to isolate clean spectra by imposing γ-ray coincidences. Clean electron lines could only be obtained when gates were placed on coincident γ rays. In particular, gating on the intense yrast transitions could, in many cases, remove the contaminating yrare transitions. Figure 3 shows the electron and γ-ray spectra when gated by the $6^+ \rightarrow 4^+$, 351-keV γ ray of the ground-state band.

Conversion coefficients were obtained for a total of 21 transitions in ^{182}W. The remaining transitions were either too weak, or could not be sufficiently resolved to extract

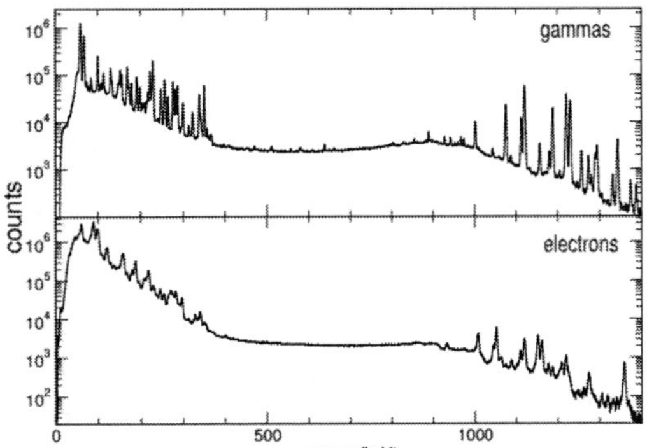

Figure 2. Total projection of γ rays (top) and electrons (bottom) in ^{182}W. Spectra are binned at 0.5 keV/channel

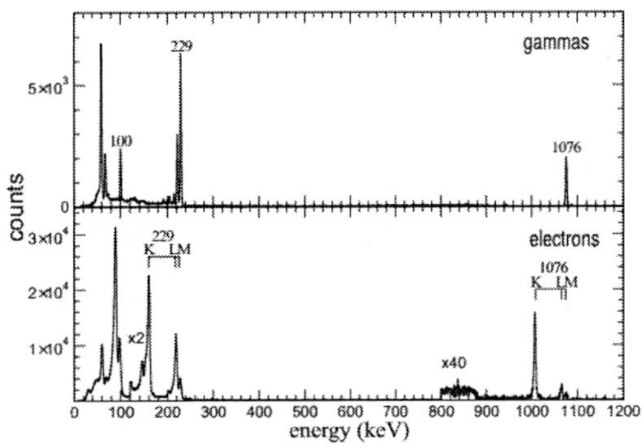

Figure 3. Gamma rays (top) and electrons (bottom) coincident with the 351-keV γ ray. Spectra are binned at 0.5 keV/channel

conversion coefficients. The results of the conversion coefficient measurements were compared to theoretical values that were calculated using BrIcc [13]. In general there was good agreement between the measured and theoretical values (see Fig. 4).

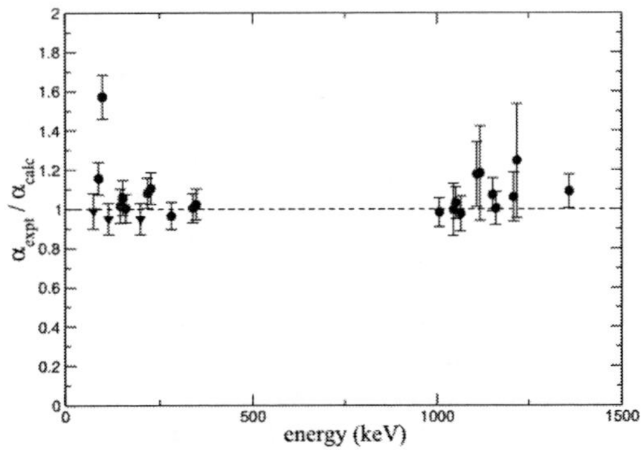

Figure 4. Comparison of measured and theoretical conversion coefficients for ^{182}W (circles) and ^{152}Sm (triangles). Data for ^{152}Sm was obtained with an open calibration source of ^{152}Eu.

Three of the transitions measured were known to have $E0$ admixtures and their $E0/E2$ mixing ratios (q_k) were extracted (see Table 1). The $E0/E2$ mixing ratios evaluated in the current work are in agreement with those previously reported in Ref. [10], however, the large uncertainty means new information cannot be extracted. Nevertheless, the results show the potential power of the Solenogam system.

Table 1. Comparison of $E0/E2$ mixing ratios measured in the current work with values from Ref. [10].

Transition	q_k	
E_γ (keV)	Expt.	Ref. [10]
1076	<0.2	0
1113	<0.4	0.41 (9)
1121	0.21 (16)	0.16 (9)

2.3 Angular correlations

The angular differences between the HPGe and Si(Li) detectors in the Solenogam array and corresponding detector pairs for the current work are given in Table 2.

Table 2. Angular differences and corresponding detector pairs in Solenogam. The notation γ_i and e_j^- refers to the different HPGe and Si(Li) detectors in the array, respectively, and θ_{ij} refers to the angular difference between the detectors.

θ_{ij}	$\cos^2\theta_{ij}$	Detector pairs			
90°	0.0	γ_1, e_3^-	γ_2, e_2^-	γ_3, e_1^-	γ_3, e_4^-
104.5°	0.063	γ_1, e_2^-	γ_1, e_6^-	γ_2, e_1^-	γ_2, e_3^-
120°	0.25	γ_3, e_2^-	γ_3, e_5^-		
138.6°	0.563	γ_1, e_1^-	γ_1, e_5^-	γ_2, e_4^-	γ_2, e_6^-
159°	0.87	γ_3, e_3^-	γ_3, e_6^-		
180°	1.0	γ_1, e_4^-	γ_2, e_5^-		

Angular correlation data was obtained from γ-e$^-$ coincidence matrices constructed for each angular difference in the Solenogam array. Each matrix contained events only from γ-ray and electron detector pairs separated by the corresponding angle. As the array was used with only three HPGe detectors, there were insufficient angular differences between γ-ray detectors to make γ-γ correlation measurements feasible. The γ-e$^-$ angular correlation intensities were extracted by gating on the populating γ-ray in the cascade and measuring the intensity of the depopulating electron and then a Legendre polynomial fit was used to extract A_2 and A_4 values. Typical results are shown in Fig. 5.

Gamma-electron angular correlations were observed for seven of the strongest cascades in ^{182}W (see Table 3). Other cascades were too weak for correlation data to be extracted. The expected angular correlations were calculated using particle parameters b_2 and b_4 from Ref. [14], and mixing ratios from Ref. [9]. There was qualitative agreement between the fitted and the calculated correlations (see Fig. 5), however, Fig. 6 shows that the fitted A_2 values where consistently 40% less then the calculated values. The large uncertainty associated with the fitted A_4 values meant that they where indistinguishable from zero.

Table 3. Angular correlations measured in the current work.

Cascade	Energy (keV)	States involved
1	229-100	$4 \to 2 \to 0$
2	1121-100	$2 \to 2 \to 0$
3	1342-100	$4 \to 2 \to 0$
4	351-229	$6 \to 4 \to 2$
5	1427-229	$6 \to 4 \to 2$
6	1076-229	$6 \to 6 \to 4 \to 2$
7	1076-351	$6 \to 6 \to 4$

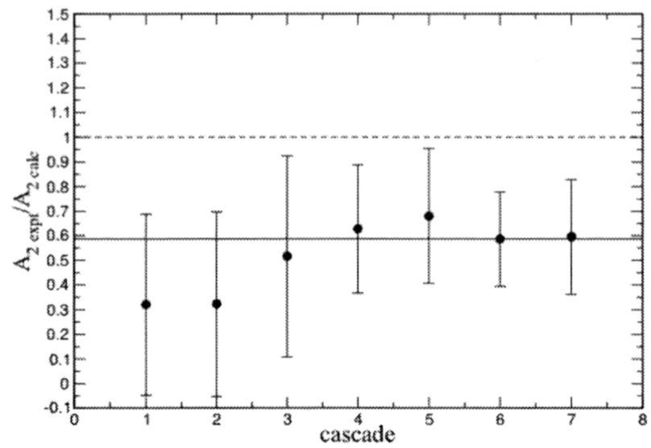

Figure 6. Comparison of measured and theoretical A_2 values. The dashed line at 1 indicates agreement between fitted and calculated values. The solid line at 0.56(10) indicates the weighted average.

The conversion coefficients were first evaluated without correcting for the angular correlation effects, however, the already good agreement with theoretical calculations suggests that the addition of correlations into the evaluation should have minimal effect. Indeed, it had no noticeable effect on the measured conversion coefficients. This result is unsurprising as the wide range of $\cos^2(\theta)$ values sampled in the Solenogam array is expected to average out the correlations. It should be noted that the theoretical correlations were used for these corrections and the observed attenuation was not considered. While this attenuation is still unexplained, it further reduces the impact of the angular correlations on the conversion coefficients.

3 Discussion

The present work has developed Solenogam by using a ^{182}Re source to characterise the performance of the system in obtaining γ-e$^-$ coincidences. These coincidences have been used to study conversion coefficients and γ-e$^-$ angular correlations for selected transitions in ^{182}W. The $E0/E2$ mixing ratios for three $J \to J$ transitions were evaluated, showing the potential power of the system to probe nuclei that exhibit shape coexistence. The values measured in the current work agree with those previously reported, although, the significant uncertainty means no new information can be extracted. This uncertainty has been attributed to a lack of statistics in the gated-electron spectra, an issue that can be addressed in in-beam experi-

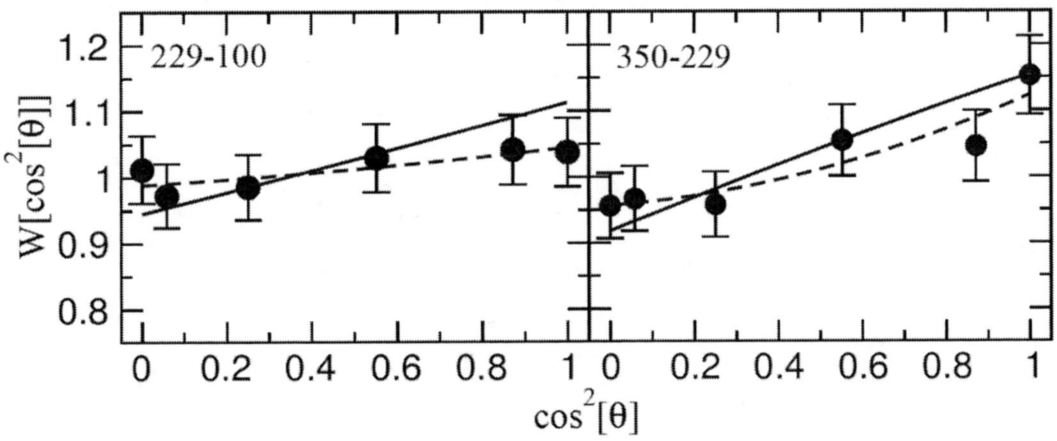

Figure 5. Selected calculated (solid) and observed (dashed) angular correlations in ^{182}W. Cascades are labelled with the energy of the populating γ ray followed by the transition energy for which the conversion electron was observed.

ments where nuclei are continuously produced, in comparison with the present offline source measurement. Of note is that while the $E2/M1$ mixing ratios used in these evaluations were taken from previously reported values, the Solenogam array is able to determine both δ and q_k values. For intense transitions, conversion coefficients can be determined for the K, L and M orbitals, and total conversion coefficients can be independently obtained from γ-γ intensity balances. This will provide ample data to uniquely determine both $E2/M1$ and $E0/E2$ mixing ratios.

The limiting factor for the extraction of conversion coefficients was a lack of suitable γ-ray gates. Many of the yrare transitions could not be isolated by gating and so clean electron spectra were not available. This could be mitigated by requiring triple coincidences, γ-γ-e$^-$ or any such permutation, however, the current setup has a small γ-ray detection efficiency for effective measurement of high-fold events.

Conversion coefficients were measured for transitions with intensity branches as low as 1.56% for low-energy (214 keV) transitions and 4.6% at higher energies (1113 keV). This trend is not unexpected as the probability for internal conversion is less at higher energies leading to weaker electron lines. For L and M electrons, the weakest low-energy transition for which conversion coefficients could be extracted was the 351-keV transition produced in 11.7% of decays. The small uncertainty associated with this value indicates that this is by no means the limit of the system. At higher energies, the 17.4%, 1221-keV transition and the 10.7%, 1076-keV transition were the limits for measuring L and M lines, respectively. However, unlike the 321-keV transition, the large uncertainty associated with the M line of the 1221-keV transition (~20%) indicates that this is approaching the limits of the array in this energy regime.

Solenogam is currently awaiting installation of the 8-T solenoid. Once this process is complete, the system can be used in-beam. This will allow access to nuclei produced using heavy-ion fusion-evaporation reactions. The Solenogam array was originally designed to study shape coexistence in nuclei in the mercury-lead region where neutron-deficient nuclei are only accessible in-beam, however, with further development other areas can be investigated. Indeed, plans are under-way to install six $LaBr_3$(Ce) detectors to enable fast-timing experiments to also be performed with the system.

4 Conclusions

The Solenogam array was used to measure conversion coefficients using γ-e$^-$ coincidences for a range of transitions in ^{182}W and good agreement was achieved with theoretical values. Qualitatively correct angular correlations were observed for a number of cascades, however, these anisotropies were attenuated by approximately 40% for reasons unclear at present. It was further verified that corrections for angular correlation effects were not significant for measuring conversion coefficients. The measurements made in the current work indicate that the Solenogam array has the potential to provide access to a range of interesting nuclei that can be studied in a unique way.

Acknowledgements

This research was supported by the Australian Research Council through grant numbers FT100100991, DP120101417, and DP140103317.

References

[1] K. Heyde and J. L. Wood, Rev. Mod. Phys. **83**, 1467 (2011).

[2] G. D. Dracoulis and A. P. Byrne, Australian National University, Department of Nuclear Physics Annual Report, **ANU-P/1052**, 115 (1989).

[3] T. Kibédi, G. D. Dracoulis and A. P. Byrne, Nucl. Instr. Meth. A **294**, 523 (1990).
[4] J. K. Jabber and N. M. Stewart, J. Phys. G **16**, 271 (1990).
[5] D. H. White and R. E. Birkett, Nucl. Phys. **A136**, 657 (1969).
[6] P. H. Regan *et al.*, Nucl. Phys. **A567**, 414 (1994).
[7] J. J. Sapyta, E. G. Funk, and J. W. Mihelich, Nucl. Phys. **A139**, 161 (1969).
[8] M. S. S. El-Daghmah, N. M. Stewart, Z. Phys. A **309**, 219 (1983).
[9] B. Singh, and J. C. Roediger, Nucl. Data Sheets **111** 2081 (2010).
[10] L. Westerberg, L. O. Edvardson, and G. Ch. Madueme, Nucl. Phys. **A255**, 427 (1975).
[11] B. D. Jeltema *et al.*, Nucl. Phys. **A280**, 21 (1977).
[12] G. P. Williams, X-Ray data booklet: Electron binding energies (2000).
[13] T. Kibédi *et al.*, Nucl. Inst. Meth. A **589**, 202 (2008).
[14] R. S. Hager, and E. C. Seltzer, Nucl. Data Sheets. A **4**, 419 (1968).

AUTHOR INDEX

AUTHOR INDEX

EDP Sciences – Web of Conferences
17, Avenue du Hoggar
Parc d'Activité de Courtabouf
BP 112
F-91944 Les Ulis Cedex A
France

ISBN 978-1-5108-3016-5

CURRAN ASSOCIATES INC.
proceedings.com

International Conference on Environment, Renewable Energy and Green Engineering (EREGCE 2022)

E3S Web of Conferences Volume 350 (2022)

Online
22-24 April 2022

edp sciences